Asymptotics and Extrapolation

Guido Walz

MATHEMATICAL
RESEARCH

Volume 88

Asymptotics and Extrapolation

Guido Walz

Akademie Verlag

Author:
PD Dr. Guido Walz, University of Mannheim

With 22 tables

1st edition
Editorial Director: Dipl.-Math. Gesine Reiher

Library of Congress Card Number pending

Die Deutsche Bibliothek – CIP-Einheitsaufnahme

Walz, Guido:
Asymptotics and extrapolation / Guido Walz. –
Berlin : Akad. Verl., 1996
(Mathematical research ; Vol. 88)
ISBN 3-05-501732-3
NE: GT

ISSN 0138-3019

Akademie Verlag is a member of the VCH Publishing Group.

Printed on non-acid paper.
The paper used corresponds to both the U.S. standard ANSI Z.39.48 – 1984 and the European standard ISO TC 46.

Printing: GAM Media GmbH, Berlin
Bookbinding: Verlagsbuchbinderei Mikolai GmbH, Berlin

Printed in the Federal Republic of Germany

Akademie Verlag GmbH	VCH Publishers, Inc.
Postfach · D-13162 Berlin	220 East 23rd Street
Federal Republic of Germany	New York, NY 10010-4606

So eine Arbeit wird eigentlich nie fertig, man muß sie für fertig erklären, wenn man nach Zeit und Umständen das möglichste getan hat.

J.W. von Goethe

Our age is proud of machines which think, but mistrusts persons who try to.

H. Mumford Jones

In der numerischen Mathematik sind nicht immer die kunstvollen mathematischen Konstruktionen erfolgreich, sondern eher alte elementare und sehr direkte Methoden. Um letztere aber theoretisch erschöpfend durchleuchten zu können, müssen mitunter sehr kunstvolle mathematische Theorien eingesetzt werden.

E. Stiefel

Preface

Some years ago, I finished my doctoral thesis, which was roughly spoken concerned with the same topic as this book, and then wrote some subsequent papers on related problems. Much to my surprise, B.I.-Wissenschaftsverlag took notice of these publications and gave me the opportunity to write a book on these matters; I was very pleased, accepted, and what you hold in your hands now is the final result of this process. In the meantime, B.I.-Wissenschaftsverlag has ceased to exist, which turned me into one of those homeless (authors). Fortunately this book is now published by Akademie Verlag.

When I started to work on this project, I still planned to write an overview over all parts of Numerical Analysis where asymptotic expansions and extrapolation methods play an important role; but rather soon this turned out to be impossible, and so I had to skip some important topics, such as asymptotic expansions in PDE-theory and general nonlinear extrapolation procedures. May authors working in these fields kindly forgive me. However, I think that the following pages still contain a nice compendium on most aspects of the present state-of-the-art in asymptotic expansion and linear extrapolation theory. I am convinced that the reader will find at least most of the information he/she is looking for when consulting this book.

Many people have contributed to this publication, some by supplying their re- and preprints, others by sending mails and answering my questions. For all that, let me thank in particular my colleagues W. Auzinger, C. Brezinski, C. Carstensen, P. Deuflhard, D. J. Evans, K. H. Förster, E. Hairer, Ch. Lubich, J. N. Lyness, M. Meister, G. Mühlbach, N. Osada, Ph. Rabinowitz, M. Reimer, C. Schneider, L. F. Shampine, A. Sidi and F. Zeilfelder.

Very special thanks are due to my academic teacher, Professor Günter Meinardus, who always encouraged me to write this book, supplied me with a lot of useful hints, and was prepared for intensive and fruitful discussions at any time.

Also, my thanks go to Hermann Engesser (B.I.-Wissenschaftsverlag) and Gesine Reiher (Akademie Verlag) for their competent help in all editorial questions and for their patience with someone like me.

Finally, I want to thank my dear wife, for her love and patience, without which this book never would have been written, and our children, who very often had to live with a father sitting at his desk instead of playing with them (which sometimes would have been the better choice, believe me!).

At the very end of this preface, I cannot do better but quote from the nice monograph by H. J. Stetter [3]: "I ventured to write this book in English because it will be more easily read in poor English than in good German by 90% of my intended readers."

Mannheim, February 1996 Guido Walz

Table of Contents

Part II: Linear Extrapolation Methods

To Christiane, Manuela, Philipp and Julian,
since mathematics is very important,
but not the most important in my life.

Introduction

In almost every branch of Numerical Analysis one has to deal with sequences $\{\sigma_n\}$, which converge to a certain limit, say c_0, as n goes to infinity. Here, the elements σ_n can be real or complex numbers, but also vectors or matrices. In most applications, it is tacitly assumed that the approximations σ_n can be computed more or less easily by a finite number of operations, while the desired value c_0 cannot, and, in addition, that the convergence of the $\sigma_n' s$ to c_0 is rather slow (think, for example, of a discretization method for the numerical solution of a differential equation).

In such situations, there is need for a convergence acceleration method, that is, a method which transforms the original sequence $\{\sigma_n\}$ into another one which converges more quickly, but of course to the same limit value c_0. Amongst the most popular choices for this are *extrapolation methods,* which constitute one of the two central topics of this book. Now, extrapolation cannot be applied without knowing that the sequence under consideration satisfies a certain necessary condition; the condition we speak of here is nothing else but the existence of an *asymptotic expansion,* which is the second main topic on the following pages.

Since it is so important for the profound understanding of the whole theory, let us summarize this connection once more: The existence of an asymptotic expansion for some given sequence is a *necessary theoretical condition* for the applicability of an extrapolation method, which is a *numerical procedure* for convergence acceleration. So, in my eyes (but not only in mine) we have here *one* interesting theory, consisting of two parts. The organization of the present book follows precisely this philosophy: It is divided into two parts, the first one dealing with asymptotic expansions, while the second one, which naturally contains most of our numerical examples, is devoted to extrapolation methods. Moreover, since our philosophy is such that extrapolation should be an as easy as possible to handle process, we concentrate on *linear* extrapolation methods and adress the case of general nonlinear extrapolation shortly in Section 6.

It is time now for two or three specific examples. First, let us consider the sequence of numbers $\sigma_n = \sigma_n(z)$, defined for all $z \in \mathbb{C}$ and $n \in \mathbb{N}$ by

$$\sigma_n(z) := \left(1 + \frac{z}{n}\right)^n ; \qquad (0.1)$$

it is known that this sequence converges to $\exp(z)$, as n tends to infinity. But it is known as well that this convergence is rather slow (more precisely, $O(n^{-1})$), such that (0.1) does not establish a practical method for the numerical calculation of the exponential function; for example, if $z = 1$, we obtain for $n = 2^7$ still an error greater than 10^{-2}. But we shall prove later (Lemma 3.45) that the sequence defined by (0.1) possesses an asymptotic expansion (more precisely, a logarithmic asymptotic expansion, see Section 3.1), which enables us to apply a suitable extrapolation procedure. Doing this, we can reduce the error to a magnitude of 10^{-9}; this will be proved and worked out in Section 4.2.

A second example, which is well-known to every Numerical Analysist such that we only need to quote the name, is established by Romberg's method for the numerical integration of functions. Romberg's method is nothing else but a special extrapolation method (see Section 4.3) and its applicability is based on the fact that the sequence $\{T_n\}$ of trapezoidal values T_n (approximations coming from the trapezoidal rule with n subintervals) possesses for sufficiently smooth integrands an asymptotic expansion in powers of n^{-2}, see Section 3.4.

Quite another type of expansion, usually denoted as geometric asymptotic expansion, exists in the following setting: Let there be given a certain iteration function T, defined on an interval I and satisfying some mild additional assumptions, and define a sequence of numbers $\{z_n\}$ recursively by the process

$$z_{n+1} := T(z_n), \quad n = 0, 1, \dots, \tag{0.2}$$

with some $z_0 \in I$. Then this sequence possesses a geometric asymptotic expansion (see Theorems 2.2, 2.3 and 2.4). Once again, extrapolation can be applied in order to accelerate the convergence, and it turns out that the extrapolation process in this case can be viewed as a generalization of Aitken's Δ^2-process. The fact that many parts of extrapolation theory were re- and rerediscovered several times in the history is unfortunately almost characteristic for this theory and appears at several places, see the Historical Notes at the end of this book.

In all these examples we have considered sequences of numbers (and not vectors or matrices), and this will also be the case in most parts of this book; but since the extrapolation methods under consideration are linear processes, there is no difficulty at all to apply the results componentwise to vector or matrix sequences.

Finally, I think that it is only fair to notify here the existence of some other publications on extrapolation methods (but not so much on asymptotic expansions) and related topics, such as the books by C.Brezinski [1], C.Brezinski & M.Redivo Zaglia [1], J.P.Delahaye [1], G.I.Marchuk & V.V.Shaidurov [1] and J.Wimp [1]. Readers who could not find the information they were looking for in the present book are kindly invited to have a look into the ones just quoted.

Part I: Asymptotic Expansions

1. Asymptotic Systems and Expansions

1.1 Basic Definitions and Examples

Almost everywhere in this book we shall deal with questions of convergence and will therefore need a measure for the speed of this convergence. Very useful tools for this are the O–Symbols, introduced by E.LANDAU [1].

Definition 1.1: Let $f := \{f_n\}$ and $g =: \{g_n\}$ be two sequences of complex numbers. Then we write

a) $f_n = O(g_n)$, if there exists a real constant A and a number $N \in I\!N$, such that

$$|f_n| \leq A \cdot |g_n| \text{ for all } n > N ,$$

b) $f_n = o(g_n)$, if

$$\frac{|f_n|}{|g_n|} \to 0 \quad \text{for } n \to \infty ,$$

c) $f_n \sim g_n$, if

$$f_n = O(g_n) \text{ and } g_n = O(f_n) .$$

Of course, this definition can be generalized in many ways, for example we could admit any complex number as limit point, or we could take a Banach space instead of the field of complex numbers. However, the present form of Definition 1.1 is the appropriate one for our purposes; for more information about the Landau symbols we refer to N.DEBRUIJN [1], A.ERDÉLYI [1], or F.OLVER [1].

Examples. 1. Frequently used in the literature are notations like $f_n = o(1)$. This simply means that the sequence f itself tends to zero, whereas $f_n = O(1)$ implies that f is bounded as n goes to infinity.
2. Let, for $n \in \mathbb{N}$, the sequence $\{f_n\}$ be defined by

$$f_n = \exp(1/n) - 1 .$$

Then we have

$$f_n = O(1/n) . \tag{1.1}$$

Proof. Using the series expansion of the exponential function we obtain for all $n \in \mathbb{N}$, $n \geq 2$, the estimation

$$n \cdot (\exp(1/n) - 1) = \sum_{\nu=1}^{\infty} \frac{1}{\nu! n^{\nu-1}} \leq \sum_{\nu=0}^{\infty} \frac{1}{n^\nu} \leq 2 ,$$

which clearly proves the statement. □

If we have two sequences, say f^1 and f^2, which both satisfy the relation

$$f_n^i = O(g_n) , \quad i = 1, 2,$$

for some sequence g, then it follows that

$$(f^1 + f^2)_n = O(g_n)$$

Some authors use for this fact the abbreviated notation

$$O(g_n) + O(g_n) = O(g_n) . \tag{1.2}$$

Similarly, one frequently finds expressions like

$$o(g_n) = O(g_n) \tag{1.3}$$

(meaning that $f_n = o(g_n)$ implies $f_n = O(g_n)$) and so on. This is a very *dangereous* thing, because reading equations (1.2) and (1.3) from the right to the left would produce nonsense (since they are in fact no equations).

We now turn to the definition of an asymptotic system (sometimes also denoted as asymptotic scale), which will be the tool for the subsequent definition of asymptotic expansions. In the present form, it goes back to G.MEINARDUS [3], see also G.MEINARDUS & G.MERZ [2].

Definition 1.2: A system $V = \{v^{(\mu)}\}$ of vectors

$$v^{(\mu)} = (v_1^{(\mu)}, v_2^{(\mu)}, \ldots) , \quad \mu \in \mathbb{N}_0 ,$$

will be called *asymptotic system*, if the following three conditions hold:

1. $v_n^{(0)} = 1$ for all $n \in \mathbb{N}$,
2. $v_n^{(\mu)} \neq 0$ for all $\mu, n \in \mathbb{N}$,
3. $v_n^{(\mu+1)} = o(v_n^{(\mu)})$ for $n \to \infty$ for all $\mu \in \mathbb{N}_0$.

Obviously, conditions 1. and 3. imply that for all $\mu \in \mathbb{N}$ we have

$$v_n^{(\mu)} = o(1) \text{ for } n \to \infty .$$

As a first *example*, consider for all $\mu \in \mathbb{N}_0$ and $n \in \mathbb{N}$ the numbers

$$v_n^{(\mu)} := n^{-\mu} .$$

The resulting system of vectors is a special case of a so-called logarithmic asymptotic system, which we will study later on in great detail (see Chapter 3).

We are now in the position to give the definition of an asymptotic expansion, which is the central topic in the first part of this book.

Definition 1.3: Let $S := \{\sigma_n\}_{\mathbb{N}}$ be a sequence of numbers or vectors, V a given asymptotic system and m a fixed natural number. We say that the sequence S possesses an *asymptotic expansion* with respect to V of *order* m, if the exist coefficients $c_0, \ldots, c_m$, independent of n, and a number $N \in \mathbb{N}_0$, such that for all $n > N$ a representation of the form

$$\sigma_n = \sum_{\mu=0}^{m} c_\mu v_n^{(\mu)} + R_{n,m} \tag{1.4a}$$

is valid, where the residual $R_{n,m}$ satisfies

$$R_{n,m} = o(v_n^{(m)}) \text{ for } n \to \infty . \tag{1.4b}$$

If the number m can be chosen arbitrarily large, we say that the expansion is of arbitrary order and use the notation

$$\sigma_n = \sum_{\mu=0} c_\mu v_n^{(\mu)} .$$

If there is no ambiguity, we shall always combine the two relations (1.4) and use the abbreviated form

$$\sigma_n = \sum_{\mu=0}^{m} c_\mu v_n^{(\mu)} + o(v_n^{(m)}) \quad \text{for } n \to \infty , \tag{1.5}$$

or, equivalently,

$$\sigma_n = c_0 + \sum_{\mu=1}^{m} c_\mu v_n^{(\mu)} + o(v_n^{(m)}) \quad \text{for } n \to \infty .$$

It should be emphasized that we in general do not assume that the infinite series

$$c_0 v_n^{(0)} + c_1 v_n^{(1)} + \cdots$$

converges.

Lemma 1.4 (cf. K.BÖHMER [1]): *If a sequence $\{\sigma_n\}$ possesses an asymptotic expansion of order m with respect to a given asymptotic system V, then*

$$\lim_{n\to\infty}\left\{\frac{1}{v_n^{(k)}}\cdot\left(\sigma_n-\sum_{\mu=0}^{k-1}c_\mu v_n^{(\mu)}\right)\right\} = c_k \tag{1.6}$$

for $k=0,\ldots,m$.
Conversely, if Relation (1.6) holds with $c_k \neq 0$ for $k=0,\ldots,m$, then the sequence $\{\sigma_n\}$ possesses an asymptotic expansion of the type (1.5).

The proof of this little lemma is easy and left to the reader. In particular, it says that the coefficients of an asymptotic expansion – for given σ_n and V – are uniquely determined.

In the special case $k=0$, we deduce from (1.6) that

$$\lim_{n\to\infty}\sigma_n = c_0 \tag{1.7}$$

for all asymptotic expansions. Therefore we denote the coefficient c_0 as the *limit* of the asymptotic expansion (1.5).

Relation (1.7) implies one of the main applications of asymptotic expansions and extrapolation schemes, namely: The approximation of the constant c_0 by the numbers σ_n for some large n. In the next sections we shall see a lot of illustrative examples for this technique; since very little can be said on the general situation, we will mostly concentrate on two main types of asymptotic expansions, which are the geometric and the logarithmic ones.

1.2 Problems

P 1.1. Prove the following (cf. K.BÖHMER [1]): A sequence $\{\sigma_n\}$ possesses an asymptotic expansion of order m, if and only if there exists an asymptotic system V and coefficients $c_0, \ldots, c_m$, such that the relation

$$\sigma_n - \sum_{\mu=0}^{k} c_\mu v_n^{(\mu)} = o(v_n^{(k)}) \text{ for } n \to \infty$$

holds for $k = 0, \ldots, m$.

P 1.2. Let $R := \{\rho_0, \rho_1, \ldots\}$ be a set of complex numbers, and define the system $V := \{v^{(\mu)}\}$ of vectors

$$v^{(\mu)} = (1^{\rho_\mu}, 2^{\rho_\mu}, 3^{\rho_\mu}, \ldots), \quad \mu \in I\!N_0 .$$

Give necessary and sufficient conditions on the set R, such that the system V forms an asymptotic system.

2. Geometric Asymptotic Expansions

The first important type of asymptotic systems and expansions we shall deal with are the so-called geometric ones. Let us begin with the precise definition of these terms.

Definition 2.1: Let $\lambda_0, \lambda_1, \lambda_2, \ldots$ be a sequence of complex numbers with

$$1 = \lambda_0 > |\lambda_1| > |\lambda_2| > \cdots > 0 \, .$$

Then the system $G := \{v^{(\mu)}\}$ of vectors

$$v^{(\mu)} := (\lambda_\mu^1, \lambda_\mu^2, \lambda_\mu^3, \ldots) \text{ for } \mu \in I\!N_0$$

is an asymptotic system in the sense of Definition 1.2. We shall denote it as *geometric asymptotic system.*

Consequently, a sequence $\{\sigma_n\}$ possesses a *geometric asymptotic expansion* of order m, if there exist numbers $\lambda_0, \lambda_1, \ldots$ like above and an $N \in I\!N_0$, such that each σ_n for $n > N$ can be written in the form

$$\left.\begin{aligned} \sigma_n &= \sum_{\mu=0}^{m} c_\mu \lambda_\mu^n + o(\lambda_m^n) \\ &= c_0 + \sum_{\mu=1}^{m} c_\mu \lambda_\mu^n + o(\lambda_m^n) \quad \text{for } n \to \infty \, . \end{aligned}\right\} \tag{2.1}$$

Furthermore, if there exists a number λ with $1 > |\lambda| > 0$, such that $\lambda_\mu = \lambda^\mu$ for all μ, we denote resulting expansion

$$\sigma_n = c_0 + \sum_{\mu=1}^{m} c_\mu \lambda^{\mu n} + o(\lambda^{mn}) \quad \text{for } n \to \infty \tag{2.2}$$

as *special geometric asymptotic expansion.*

In the following two subsections we shall study asymptotic expansions of this type from quite different branches of mathematics. First, in Subsection 2.1 it will be completely elucidated, under which conditions on the iteration function T a sequence of numbers $\{x_n\}$, defined through an iteration process of the type

$$x_n := T(x_{n-1}), \tag{2.3}$$

possesses a geometric asymptotic expansion. Subsection 2.2 is devoted to certain matrix iterative methods, which lead to asymptotic expansions with respect to the eigenvalues of the matrix under consideration.

2.1 Asymptotic Behavior of Iteration Processes

In very many branches of Numerical Analysis there appear sequences of numbers $\{x_n\}$, which are defined through an iteration process of the type (2.3). However, for a long time there existed only heuristic and numerical results about the asymptotics of such sequences, and G. Meinardus seems to be the first who gave an exact proof of the fact that – under certain mild assumptions on the iteration function T – the resulting sequence $\{x_n\}$ possesses a special geometric asymptotic expansion. This is the first as well as the main result in the present subsection.

Theorem 2.2 (G.MEINARDUS [3,4]): *Consider a real function T, defined on some closed interval I and having the properties*

1. $T(I) \subset I$,
2. $T \in C^{m+1}(I)$ *and*
3. $|T'(x)| \leq \rho < 1$ *for all* $x \in I$,

where ρ is a constant and m some fixed integer.
Furthermore, denote the unique fixed point of T in I by ξ, and define for arbitrary $x_0 \in I$ a sequence of real numbers $\{x_n\}$ through $x_{n+1} := T(x_n)$.
Then the sequence $\{x_n\}$ possesses the special geometric asymptotic expansion of order m

$$x_n = \xi + \sum_{\mu=1}^{m} c_\mu \lambda^{\mu n} + o(\lambda^{mn}) \quad \text{for } n \to \infty \tag{2.4}$$

with $\lambda := T'(\xi)$, provided that this value is different from zero.

Proof. For $x \in I$ we can expand the function $T(x)$ around ξ and obtain

$$T(x) = \sum_{j=0}^{m} \frac{T^{(j)}(\xi)}{j!}(x-\xi)^j + O((x-\xi)^{m+1}).$$

Putting $x := x_n$ yields

$$x_{n+1} = \xi + \sum_{j=1}^{m} \frac{T^{(j)}(\xi)}{j!}(x_n-\xi)^j + O((x_n-\xi)^{m+1}), \tag{2.5}$$

since $T(\xi) = \xi$. Equation (2.5) shows that for each n the approximation error $e_n := x_n - \xi$ can be written in the form

$$e_{n+1} = \lambda e_n + \sum_{j=2}^{m} \gamma_j e_n^j + O(e_n^{m+1}), \tag{2.6}$$

where the constants γ_j , $j = 2, \ldots, m$, do not depend on n .
Now, (2.6) is a nonlinear difference equation of first order for the sequence $\{e_n\}$; as is known from the theory of difference equations (cf. J.HORN [1] or T.L.SAATY [1]), this equation has a solution of the form

$$e_n = \sum_{\mu=1}^{m} c_\mu k^{\mu n} + o(k^{mn}) \quad \text{for } n \to \infty , \tag{2.7}$$

with some real numbers c_μ , $\mu = 1, \ldots, m$, $c_1 \neq 0$, and k , which do not depend on the natural number n .

Obviously $|k| < 1$, since the sequence $\{x_n\}$ is already known to converge, hence $\lim_{n\to\infty} e_n = 0$. So (2.7) would prove the assertion of the theorem, if we could show that $k = \lambda$. To do this, we study the behaviour of the quotient e_{n+1}/e_n , as n goes to infinity. From (2.6) it follows that

$$\lim_{n\to\infty} \frac{e_{n+1}}{e_n} = \lambda . \tag{2.8}$$

On the other hand it follows from (2.7) that

$$\frac{e_{n+1}}{e_n} = \frac{c_1 k^{n+1} + O(k^{2n+2})}{c_1 k^n + O(k^{2n})} \quad \text{for } n \to \infty ,$$

which implies

$$\lim_{n\to\infty} \frac{e_{n+1}}{e_n} = k . \tag{2.9}$$

Comparing (2.8) and (2.9), we recognize that (2.7) is identical with (2.4), which proves the assertion of Theorem 2.2. □

Example. With a sufficiently smooth real function f , possessing in $\xi \in I\!R$ a zero of multiplicity $k \in I\!N$, we consider *Newton's method* for calculating roots, i.e. we use the iteration function

$$T(x) = x - \frac{f(x)}{f'(x)} . \tag{2.10}$$

In this case, the value of $T'(\xi)$ is independent of ξ and can be computed a priori, it is $T'(\xi) = \frac{k-1}{k}$. So, for $k \geq 2$ the assumptions of Theorem 2.2 are satisfied in a neighborhood I of ξ, and we obtain the following result:
The sequence $\{x_n\}$, defined through the iterative process $x_{n+1} := T(x_n)$ with T from (2.10), possesses the asymptotic expansion

$$x_n = \xi + \sum_{\mu=1}^{m} c_\mu \left(\frac{k-1}{k}\right)^{\mu n} + o\left(\left(\frac{k-1}{k}\right)^{mn}\right) \quad \text{for } n \to \infty ,$$

where the order m depends on the smoothness of f. □

In Theorem 2.6, we shall present a generalization of Theorem 2.2 to multivariate functions, which is due to H. Ortloff [1,2]. However, there in general the resulting asymptotic expansion is no longer of geometric type.

In the case of a complex sequence, a result quite similar to that of Theorem 2.2 is valid:

Theorem 2.3 (G. Meinardus [4]): *Let M be a bounded, closed and convex subset of a region G in the complex plane, and consider a complex-valued function T, which is holomorphic on G and has the properties*

1. $T(M) \subset M$,
2. $\sup_{z \in M} |T'(z)| \leq \rho$,

with some constant $\rho < 1$.
Furthermore, denote the unique fixed point of T in M by ξ, and define for arbitrary $z_0 \in M$ a sequence of complex numbers $\{z_n\}$ through $z_{n+1} := T(z_n)$.
Then the sequence $\{z_n\}$ possesses the special geometric asymptotic expansion

$$z_n = \xi + \sum_{\mu=1} c_\mu \lambda^{\mu n} \quad \textit{for } n \to \infty \tag{2.11}$$

of arbitrary order, provided that the value $\lambda := T'(\xi)$ *is different from zero.*

The *proof* of Theorem 2.3 conincides up to a certain degree with the original one of Theorem 2.2; we refer to the original paper by G.MEINARDUS [4]. □

The most restrictive assumption of Theorem 2.2 (resp. Theorem 2.3) concerning the iteration function T is surely that $T'(\xi)$ must not vanish, and there is great interest in the question, what happens if this assumption is dropped. Of course it is well-known that we will have no longer linear convergence in this case, i.e. an asymptotic expansion of the type (2.4) cannot exist, but since Meinardus' original proof of Theorem 2.2 cannot be modified (at least not in an obvious way) to the case $T'(\xi) = 0$, this question remained open for some time.

However, our proof from above can indeed be modified, and the corresponding result is given in the following Theorem 2.4. In combination with Theorem 2.2, this answers the question for asymptotic expansions of iteratively defined real sequences in a quite satifactory way.

Theorem 2.4 (G.WALZ [10]): *Let the real function* T *be defined on an interval* I *and belong to the class* $C^{s+1}(I)$ *for some* $s \in \mathbb{N}$, $s \geq 2$. *Furthermore, denote by* ξ *an isolated fixed point of* T *in* I *and suppose that*

$$T'(\xi) \;=\; \cdots \;=\; T^{(p-1)}(\xi) \;=\; 0\,,\; T^{(p)}(\xi) \neq 0$$

for some $p \in \mathbb{N}$ *with* $2 \leq p \leq s$.
Then there exists a neighborhood $U_\varepsilon(\xi)$ *of* ξ *such that for each* $x_0 \in U_\varepsilon(\xi)$ *the sequence* $\{x_n\}$, *defined by* $x_{n+1} := T(x_n)$, *possesses the asymptotic expansion*

$$x_n \;=\; \xi + \sum_{\mu=1}^{r} c_\mu \lambda^{\mu p^n} + o(\lambda^{r p^n}) \;\; \text{for } n \to \infty \qquad (2.12)$$

with some $r = r(s,p) \in \mathbb{N}$ *and* $\lambda \in \mathbb{R}$, $|\lambda| < 1$.
The asymptotic constant c_1 *can here be computed explicitly; it is given by*

$$c_1 = \left(\frac{p!}{T^{(p)}(\xi)}\right)^{\frac{1}{p-1}} . \tag{2.13}$$

Proof. Proceeding in analogy to the proof of Theorem 2.2, we see that the sequence of errors e_n now satisfies a relation of the form

$$e_{n+1} = \sum_{j=p}^{s} \gamma_j e_n^j + o(e_n^s) \quad \text{for } n \to \infty \tag{2.14}$$

with $\gamma_j = T^{(j)}(\xi)/j!$ for $j = p, \ldots, s$. This is again a nonlinear difference equation of first order, and we can conclude from T.L.SAATY [1, p.179] that the solution of (2.14) is of the form

$$e_n = \sum_{\mu=1}^{r} c_\mu \lambda^{\mu p^n} + o(\lambda^{r p^n}) \quad \text{for } n \to \infty \tag{2.15}$$

with some natural number r, depending on s and p, and some real number λ, depending on e_0, but not on n.
Now, choose ε small enough, such that the sequence $\{x_n\}$ converges to ξ for all $x_0 \in U_\varepsilon(\xi)$; then of course $\lim_{n\to\infty} e_n = 0$ for all $e_0 \in U_\varepsilon(0)$, and therefore $|\lambda| < 1$ as a consequence of (2.15). This proves the first assertion of the theorem.

In order to show (2.13) we analyze the limit behavior of the quotient e_{n+1}/e_n^p for $n \to \infty$. From (2.14) we obtain

$$\lim_{n\to\infty} \frac{e_{n+1}}{e_n^p} = \frac{T^{(p)}(\xi)}{p!} , \tag{2.16}$$

whereas (2.15) implies

$$\lim_{n\to\infty} \frac{e_{n+1}}{e_n^p} = c_1^{1-p} . \tag{2.17}$$

Now the combination of equations (2.16) and (2.17) implies the assertion (2.13) □

It must be pointed out that the asymptotic expansion (2.12) is *not* a geometric one. Maybe one could denote it as *supergeometric* (or *hypergeometric)* asymptotic expansion for the moment.

Let us examine a little *example* illustrating Theorem 2.4; for some fixed real number $a > 0$ we consider Newton's method for computing the square root of a, i.e. the iteration function

$$T(x) = \frac{1}{2}\left(x + \frac{a}{x}\right) \quad \text{for } x > 0$$

with the unique fixed point $\xi = \sqrt{a}$. Here

$$\begin{aligned} T'(\xi) &= 0 \text{ and} \\ T''(\xi) &= \frac{1}{\sqrt{a}} > 0\,, \end{aligned}$$

so that we are in the situation of Theorem 2.4 with $s = \infty$ and $p = 2$; thus the iteratively defined sequence $\{x_n\}$, where $x_0 > 0$ and

$$x_{n+1} = \frac{1}{2}\left(x_n + \frac{a}{x_n}\right), \tag{2.18}$$

possesses an asymptotic expansion of the type given in (2.12). Even more, a little computation shows that each x_n has the explicit representation (cf. G.MEINARDUS [3])

$$x_n = \sqrt{a}\cdot\frac{1+\lambda^{2^n}}{1-\lambda^{2^n}} \tag{2.19}$$

with

$$\lambda = \frac{x_0 - \sqrt{a}}{x_0 + \sqrt{a}}\,, \tag{2.20}$$

which shows that in this case λ depends strongly on the initial value x_0.

Expanding (2.19) into a geometric series, we obtain the representation

$$x_n = \sqrt{a} + \sum_{\mu=1} c_\mu \lambda^{\mu 2^n}$$

with λ from (2.20) and $c_\mu = 2\sqrt{a}$ for all μ, in full accordance with the assertions of Theorem 2.4.

2.2 Eigenvalue Expansions

Geometric asymptotic expansions, which are not of the special type (2.2), appear typically in the context of matrix iterative processes. In this subsection we shall study two illustrative cases, which cover almost all that is known about this topic.

2.2.1 One-Step Matrix Iteration. Consider the following situation: Given a complex $(p \times p)-$ matrix A and a vector $b \in \mathbb{C}^p$, we search for a solution of the linear system of equations $x = Ax + b$, i.e. for a vector $\xi \in \mathbb{C}^p$ with the property

$$\xi = A\xi + b . \tag{2.21}$$

Under certain assumptions this solution exists, is unique, and can be computed as limit of the vector sequence $\{x^{[n]}\}$, which is iteratively defined through

$$x^{[n]} = Ax^{[n-1]} + b \text{ with some } x^{[0]} \in \mathbb{C}^p . \tag{2.22}$$

The next theorem says that, under certain conditions on A, the sequence $\{x^{[n]}\}$ posseses a finite asymptotic expansion with limit ξ; it can be viewed as a special case of a result due to A.Sidi & J.Bridger [1], which itself is covered by Theorem 2.6.

Theorem 2.5 (A.Sidi [2]): *Consider a complex $(p \times p)-$ matrix A, possessing p linearly independent eigenvectors, and assume that the spectral radius ρ of A satisfies $\rho < 1$.*
Then equation (2.21) has a unique solution ξ, and the sequence of vectors $\{x^{[n]}\}$, defined by the iterative process (2.22), possesses the finite geometric asymptotic expansion

$$x^{[n]} = \xi + \sum_{\mu=1}^{k} c_\mu \lambda_\mu^n \tag{2.23}$$

with $k \leq p$ and some vectors $c_1, \ldots, c_k$.
Here, $\lambda_1, \ldots, \lambda_k$ are the non-zero eigenvalues of A.

Proof. The existence and unicity of the solution ξ follows directly from the assumption $\rho < 1$.
Furthermore, since the eigenvectors $v_1, \ldots, v_p$ of A provide a basis of $\mathbb{C}^p$, there exist coefficients $a_1, \ldots, a_p$, so that the vector $x^{[0]} - \xi$ can be written as

$$x^{[0]} - \xi = \sum_{\mu=1}^{p} a_\mu v_\mu . \tag{2.24}$$

Combining (2.21) and (2.22), we obtain the relation

$$x^{[n]} - \xi = A^n (x^{[0]} - \xi) , \tag{2.25}$$

which is valid for all $n \in \mathbb{N}$, and substituting (2.24) into (2.25) yields

$$\begin{aligned} x^{[n]} - \xi &= A^n \cdot \sum_{\mu=1}^{p} a_\mu v_\mu = \sum_{\mu=1}^{p} a_\mu A^n v_\mu \\ &= \sum_{\mu=1}^{p} a_\mu \lambda_\mu^n v_\mu = \sum_{\mu=1}^{k} a_\mu \lambda_\mu^n v_\mu . \end{aligned}$$

Finally, putting $c_\mu := a_\mu v_\mu$ for all μ completes the proof. □

The most general result on the existence of asymptotic expansions for matrix iterative methods is probably the following theorem by H. Ortloff. However, as already pointed out, here the nice geometricity of the resulting asymptotic expansion is no longer given.

Consider some simply-connected region $D \subset \mathbb{R}^k$, $1 \leq k < \infty$, and a sequence of vectors $\{x^{[n]}\} \in D$, defined through the iterative process

$$x^{[n+1]} := T(x^{[n]}) \tag{2.26}$$

with some function $T : D \to D$. Then the following theorem holds.

Theorem 2.6 (H.Ortloff [1,2]): *Let the iteration function T belong to the class $C^{m+1}(D)$ and possess a unique fixed point $\xi \in D$. Furthermore, assume that the matrix $T'(\xi)$ is regular and that its spectral radius ρ satisfies $\rho < 1$.*
Then the sequence of vectors $\{x^{[n]}\}$, defined through the iterative process (2.26), converges for all $x^{[0]}$ from a sufficiently small neighborhood of ξ ; in particular, the asymptotic expansion

$$x^{[n]} = \xi + \sum_{i=1}^{q} \sum_{j=0}^{p_i - 1} c_{ij} n^j \mu_i^n + o((\rho + \varepsilon)^{(m+1)n}) \tag{2.27}$$

with $\varepsilon > 0$ holds. Here, the μ_i 's are the eigenvalues of $T'(\xi)$ and certain products of them, p_i is the corresponding multiplicity, and q some finite natural number.

2.2.2 The Power Method. This method – sometimes also denoted as von Mises' method – is a well-known procedure for the numerical computation of eigenvectors and eigenvalues of certain matrices. In the sequel we shall improve the usual error estimates for this method by proving that the approximating sequence, which the method defines, possesses a geometric asymptotic expansion with limit z , the desired eigenvector. As preparation for this we need the following lemma.

Lemma 2.7: *Let $\{\sigma_n^1\}$ and $\{\sigma_n^2\}$ be two sequences which possess geometric asymptotic expansions of order m with respect to the same asymptotic system G , i.e.*

$$\sigma_n^j = \sum_{\mu=0}^{m} c_\mu^j \lambda_\mu^n + o(\lambda_m^n) \quad \textit{for } n \to \infty \tag{2.28}$$

for $j = 1, 2$, *with certain coefficients* c_μ^j, $j = 1, 2, \mu = 0, \ldots, m$, *and* $c_0^2 \neq 0$.
Then the sequence of quotients σ_n^1 / σ_n^2 *possesses also a geometric asymptotic expansion, the limit of which is the quotient* c_0^1 / c_0^2.

Proof. Let n be large enough; then $\sigma_n^2 \neq 0$, and we obtain

$$\frac{\sigma_n^1}{\sigma_n^2} = \frac{\sum\limits_{\mu=0}^{m} c_\mu^1 \lambda_\mu^n \; + \; o(\lambda_m^n)}{\sum\limits_{\mu=0}^{m} c_\mu^2 \lambda_\mu^n \; + \; o(\lambda_m^n)}$$

$$= \left\{ \frac{c_0^1}{c_0^2} + \sum_{\mu=1}^{m} \frac{c_\mu^1}{c_0^2} \cdot \lambda_\mu^n \; + \; o(\lambda_m^n) \right\} \cdot \frac{1}{1 + \sum\limits_{\mu=1}^{m} \frac{c_\mu^2}{c_0^2} \cdot \lambda_\mu^n \; + \; o(\lambda_m^n)}$$

$$= \left\{ \frac{c_0^1}{c_0^2} + \sum_{\mu=1}^{m} \frac{c_\mu^1}{c_0^2} \cdot \lambda_\mu^n \; + \; o(\lambda_m^n) \right\} \cdot$$

$$\cdot \sum_{\nu=0}^{\infty} (-1)^\nu \left(\sum_{\mu=1}^{m} \frac{c_\mu^2}{c_0^2} \cdot \lambda_\mu^n \; + \; o(\lambda_m^n) \right)^\nu .$$

Multiplying out, we see that the right hand side can be written in the form

$$\frac{c_0^1}{c_0^2} + \sum_{\nu=1} \sum_{\substack{p_1 + \cdots p_m = \nu \\ p_1, \ldots, p_m \in N_0}} c_{p_1, \ldots, p_m, \nu} (\lambda_1^{p_1} \cdots \lambda_m^{p_m})^n + o(\lambda_m^n) ,$$

where the summation runs over all products $\lambda_1^{p_1} \cdots \lambda_m^{p_m}$ with

$$|\lambda_1^{p_1} \cdots \lambda_m^{p_m}| \; \geq \; |\lambda_m| .$$

This proves Lemma 2.7. □

The following theorem improves the usual error analysis of the power method; readers not so familiar with this method can be referred to almost any textbook on Numerical Analysis, e.g. G.Meinardus & G.Merz [2].

Theorem 2.8 (cf. J.W.SCHMIDT [1], G.WALZ [1], H.WERNER [1]): *Let A be a diagonalizable $(p \times p)$-matrix with eigenvalues $l_1, \ldots, l_p$, satisfying $l_1 > 0$ and*

$$l_1 = \cdots = l_k > |l_{k+1}| \geq \cdots \geq |l_p| \tag{2.29}$$

with some $k \in \{1, \ldots, p\}$.
Furthermore, suppose that the corresponding eigenvectors $z_1, \ldots, z_p$ are linearly independent, and take a fixed vector $x^{[0]} \in \mathbb{C}^p$, such that in the decomposition

$$x^{[0]} = \sum_{j=1}^{p} \gamma_j z_j$$

not all coefficients $\gamma_1, \ldots, \gamma_k$ vanish. Then the following assertions hold:
1. For all $n \in \mathbb{N}$, the vectors $x^{[n]}$ and $y^{[n]}$ are well-defined by the iteration rule

$$\left.\begin{aligned} y^{[n]} &:= A \cdot x^{[n-1]} , \\ x^{[n]} &:= \frac{y^{[n]}}{\|y^{[n]}\|} , \end{aligned}\right\} \tag{2.30}$$

where $\|\cdot\|$ denotes the maximum norm in $\mathbb{C}^p$, and the sequence of vectors $\{x^{[n]}\}$ converges to an eigenvector z of A, which corresponds to l_1 and is normalized by $\|z\| = 1$.
2. If, for all n large enough, the norm of all vectors $x^{[n]}$ is determined by the same component, then the sequence of these vectors possesses a geometric asymptotic expansion with limit z, i.e. we can write each $x^{[n]}$ in the form

$$x^{[n]} = z + \sum_{\mu=1} \lambda_\mu^n e_\mu .$$

Here, $e_1, e_2, \ldots$ are some coefficient vectors independent of n, and $\lambda_1, \lambda_2, \ldots$ denote certain products of factors of the form l_j/l_1, $j \geq k+1$; for example, it is

$$\lambda_1 = \frac{l_{k+1}}{l_1} \quad \textit{and} \quad \lambda_2 = \max\{\frac{l_{k+1}^2}{l_1^2}, \frac{l_{k+2}}{l_1}\} .$$

Proof. The first assertion is proved in G.MEINARDUS & G.MERZ [2], where one also can find the representation

$$x^{[n]} = \frac{\sum_{j=1}^{p} \gamma_j l_j^n z_j}{\| \sum_{j=1}^{p} \gamma_j l_j^n z_j \|}$$

$$= \frac{z + \sum_{j=k+1}^{p} \gamma_j \left(\frac{l_j}{l_1}\right)^n z_j}{\| z + \sum_{j=k+1}^{p} \gamma_j \left(\frac{l_j}{l_1}\right)^n z_j \|},$$

since $l_1 \neq 0$. Now let n be large enough and denote by ζ resp. ζ_j that components of z resp. z_j, such that

$$\| z + \sum_{j=k+1}^{p} \gamma_j \left(\frac{l_j}{l_1}\right)^n z_j \| = | \zeta + \sum_{j=k+1}^{p} \gamma_j \left(\frac{l_j}{l_1}\right)^n \zeta_j | .$$

Then

$$x^{[n]} = \frac{z + \sum_{j=k+1}^{p} \gamma_j \left(\frac{l_j}{l_1}\right)^n z_j}{| \zeta + \sum_{j=k+1}^{p} \gamma_j \left(\frac{l_j}{l_1}\right)^n \zeta_j |} . \tag{2.31}$$

Using the fact that the denominator sequence in (2.31) has an asymptotic expansion of the desired type (cf. Problem P 2.1), the application of Lemma 2.7 to each component of the vector in (2.31) provides the desired result. □

From (2.30) it is clear that the sequence $\{y^{[n]}\}$ converges to the vector

$$A \cdot z = l_1 \cdot z .$$

So, if we denote the μ-th components of the vectors $x^{[n]}$ and $y^{[n]}$ by $x_\mu^{[n]}$ resp. $y_\mu^{[n]}$, we see that each one of the p sequences

$$\left\{ \frac{y_\mu^{[n]}}{x_\mu^{[n-1]}} \right\} \tag{2.32}$$

converges to the eigenvalue l_1, provided that the denominators in (2.32) are different from zero; but since we already know that $\{x^{[n]}\}$ converges to an eigenvector of A, this is clearly the case for at least one of them.

We close this subsection with a little

Example. Consider the real-symmetric matrix

$$A = \begin{pmatrix} 3 & 1 & 2 \\ 1 & 4 & 5 \\ 2 & 5 & 3 \end{pmatrix}$$

with the eigenvalues (all numbers rounded)

$$l_1 = 9.2287632378 \;, \; l_2 = 2.45111178800 \;, \; l_3 = -1.67987502585 \;;$$

the eigenvector corresponding to l_1 reads

$$z = \begin{pmatrix} 0.466377389 \\ 1.000000000 \\ 0.952477169 \end{pmatrix}$$

with $\|z\| = 1$. The first approximations for z, starting with $x^{[0]} = (1,1,1)^T$, are

$$x^{[1]} = \begin{pmatrix} 0.600000000 \\ 1.000000000 \\ 1.000000000 \end{pmatrix}, \; x^{[2]} = \begin{pmatrix} 0.500000000 \\ 1.000000000 \\ 0.958333333 \end{pmatrix},$$
$$x^{[3]} = \begin{pmatrix} 0.475336323 \\ 1.000000000 \\ 0.955156951 \end{pmatrix}, \; x^{[4]} = \begin{pmatrix} 0.468734852 \\ 1.000000000 \\ 0.952981095 \end{pmatrix}.$$

We see after a little computation that the vectors $x^{[n]}$ possess the asymptotic expansion

$$x^{[n]} = z + \sum_{\mu=1} \lambda_\mu^n e_\mu \tag{2.33a}$$

with

$$\begin{aligned}\lambda_1 &= l_2/l_1 &&= 0.26559\dots\,, \\ \lambda_2 &= l_3/l_1 &&= -0.18202\dots\,, \\ \lambda_3 &= l_2^2/l_1^2 &&= 0.07054\dots\,, \\ \lambda_4 &= l_2 l_3/l_1^2 &&= -0.04834\dots\,,\end{aligned} \tag{2.33b}$$

and so on, in accordance with Theorem 2.8 (The example will be continued in Part II).

Another numerical problem connected with the eigensytem of a matrix is that of computing the partial derivatives of the eigenvectors and eigenvalues of a parameter-dependent matrix. This problem arises in different fields of engineering, see for example R.C.Tan [1], C.Brezinski & M.Redivo Zaglia [1] and the references therein.

Interestingly, one of the most efficient methods for computing these derivatives is also based on the fact that the a certain approximating sequence possesses a geometric asymptotic expansion. Let us work this briefly out:

Consider the eigenvalue problem

$$(A - l_i I)x_i = 0 \quad \text{for } i = 1, \dots, n\,, \tag{2.34}$$

where A is a complex $(p \times p)-$ matrix and its eigenvalues l_i as well as the corresponding eigenvectors x_i depend smoothly on some real parameters, say $P_1, \dots, P_p$. Furthermore, let the eigenvalues be ordered such that

$$|l_1| \geq |l_2| \geq \cdots \geq |l_p|\,, \tag{2.35}$$

and assume that the eigenvectors are normalized by $x_i^* x_i = 1$ for all i .

Denoting respectively by $l_{i,j}$, $x_{i,j}$ and A_j the partial derivatives with respect to P_j of l_i , x_i and A , we may now define the following iterative procedure: Fix some pair of indices (i,j) and choose a vector u_0 such that $x_i^* u_0 = 0$. Then, for $n = 0, 1, \dots$, define

$$\nu_n = x_i^*(A - l_i I)u_n + x_i^* A_j x_i\,, \quad \text{and}$$

$$u_{n+1} = \{(A_j - \nu_n I)x_i + A u_n\}/\lambda_i\,.$$

This procedure was proposed by C.S.Rudisill & Y.Y.Chu [1] for the real case and later generalized by R. C. Tan to the complex case, see R.C.Tan [1] and the references therein. Moreover, it was proved by A.L.ANDREW [1,2] that if A is non-defective then for all n the following relation is true:

$$u_n = x_{i,j} + \sum_{\substack{\mu=1 \\ \mu \neq i}}^{p} (l_\mu / l_i)^n e_\mu \,. \tag{2.36}$$

Here, the e_μ 's are certain vectors, which do not depend on n .

So, if in addition to (2.35) $|l_1| > |l_2|$ holds, then, for $i = 1$ (2.36) yields a geometric asymptotic expansion in the sense of our definition. It follows that the sequence $\{u_n\}$ converges to $x_{1,j}$ (and therefore also $\{\nu_n\}$ to $l_{1,j}$), and the convergence is of order $O(|l_2/l_1|^n)$.

Moreover, it was worked out by R. C. Tan in a series of publications (see R.C.TAN [1] or C.BREZINSKI & M.REDIVO ZAGLIA [1] that also for $i > 1$ the application of suitable extrapolation methods leads to efficient algorithms for the numerical computation of the corresponding $x_{i,j}$'s and $l_{i,j}$'s.

2.3 Problems

P 2.1. Prove the following: If a sequence $\{\sigma_n\}$ of complex numbers possesses a geometric asymptotic expansion with limit c_0 , then the sequence of absolute values $\{|\sigma_n|\}$ possesses an asymptotic expansion of the same type; its limit is $|c_0|$.

P 2.2 (G.MEINARDUS [3]). Consider the complex $(2\times 2)-$ matrix

$$A := \begin{pmatrix} a & b \\ c & d \end{pmatrix} ,$$

and assume that the eigenvalues κ_1, κ_2 of A satisfy the relation $0 < |\kappa_1| < |\kappa_2|$.

Prove the following: The sequence of complex numbers $\{z_n\}$, which is defined recursively through

$$z_{n+1} = \frac{a\, z_n + b}{c\, z_n + d}$$

with some arbitrary initial value $z_1 \in \mathbb{C}$, possesses the special geometric asymptotic expansion

$$z_n = \sum_{\mu=0} c_\mu \omega^{\mu n} ,$$

where $\omega = \kappa_1/\kappa_2$.

Hint: Prove the existence of complex numbers $\alpha, \beta, \gamma, \delta$, such that

$$z_n = \frac{\alpha + \beta\omega^n}{\gamma + \delta\omega^n}$$

for all $n \in \mathbb{N}$.

P 2.3 (G.Meinardus [4], G.Walz [10]). Let the sequence $\{x_n\}$ of real numbers possess the special geometric asymptotic expansion of order m

$$x_n = \xi + \sum_{\mu=1}^{m} c_\mu \lambda^{\mu n} + o(\lambda^{mn}) \quad \text{for } n \to \infty .$$

Prove the following: For each fixed $k \in \{1, \ldots, m-1\}$, the sequence $\{q_{nk}\}$ of the real numbers

$$q_{nk} := \left(\frac{x_{n+2} - x_{n+1}}{x_{n+1} - x_n} \right)^k$$

possesses a special geometric asymptotic expansion of order m with limit λ^k, i.e.

$$q_{nk} = \lambda^k + \sum_{\mu=1}^{m} c_{\mu k} \lambda^{\mu n} + o(\lambda^{mn}) \quad \text{for } n \to \infty .$$

P 2.4. Prove Formula (2.19).

3. Logarithmic Asymptotic Expansions

Of even more importance than the geometric asymptotic expansions are probably the logarithmic ones, which we are going to study in the present section. We start with the definition of a logarithmic asymptotic system.

Definition 3.1: Let there be given a sequence of complex numbers $\{\rho_\mu\}$ with the property

$$\rho_0 = 0 \qquad \text{and} \qquad \operatorname{Re}\rho_\mu < \operatorname{Re}\rho_{\mu+1} \ \text{ for all } \mu \in I\!N_0 \ .$$

Then the system $L := \{v^{(\mu)}\}$ of vectors

$$v^{(\mu)} := (1^{-\rho_\mu},\ 2^{-\rho_\mu},\ 3^{-\rho_\mu}, \ldots) \quad \text{for } \mu \in I\!N_0$$

will be denoted as *logarithmic asymptotic system.*

It is easy to see that an logarithmic asymptotic system is indeed an asymptotic system in the sense of Definition 1.2.

Furthermore, we say that a sequence $\{\sigma_n\}$ (or that the elements σ_n) possesses a *logarithmic asymptotic expansion* of order m , if there exists a $N \in I\!N_0$, such that each σ_n for $n > N$ can be written in the form

$$\left.\begin{aligned} \sigma_n &= \sum_{\mu=0}^{m} \frac{c_\mu}{n^{\rho_\mu}} + o(n^{-Re\rho_m}) \\ &= c_0 + \sum_{\mu=1}^{m} \frac{c_\mu}{n^{\rho_\mu}} + o(n^{-Re\rho_m}) \quad \text{for } n \to \infty \,. \end{aligned}\right\} \qquad (3.1)$$

3.1 Some Introductory Examples

In this subsection we study a handful of examples for logarithmic asymptotic expansions in order to illustrate that asymptotic expansions of this type occur in very many and quite different branches of (numerical) mathematics.
Let us begin with a very, very old one.

3.1.1 Numerical Computation of π. The following construction dates back to the age of ARCHIMEDES [1]; it is well-described in many publications on numerical quadrature and related topics, see for example the papers by C.BREZINSKI [6], J.DUTKA [1], H.ENGELS [2], S.FILIPPI [1], D.C.JOYCE [1], G.MEINARDUS [3], L.F.RICHARDSON [3], or E.STIEFEL [1].

Let A_n denote the area of a regular polygon with n vertices, which is inscribed into the unit circle. By intuition it is clear that

$$\lim_{n\to\infty} A_n = \pi , \tag{3.2}$$

and this fact was used by Archimedes for his method of approximating π. He computed one by one the numbers A_6, A_{12}, A_{24}, A_{48}, and so on, until the desired accuracy was reached, and took the last value as an approximation for π. (In fact, he did even more, namely he also computed the areas of the corresponding circumscribed polygons, say U_n, and therefore obtained in each step an inclusion of the true value, i.e. $A_n < \pi < U_n$ for all n.) It is remarkable that he did not increase the number n somehow, but he precisely *doubled* it in each step, which is an interesting connection to the modern approach by extrapolation methods; we shall come back to this remark later in Part II.

Let us now analyze Relation (3.2) in more detail. First, a little computation shows that, for $n \in \mathbb{N}$,

$$A_{2n} = n \cdot \sin\left(\frac{\pi}{n}\right) , \tag{3.3}$$

and, using the series expansion of the sine function, we obtain

$$A_{2n} \;=\; \pi \;+\; \sum_{\mu=1} \frac{c_\mu}{n^{2\mu}}$$

with

$$c_\mu \;=\; (-1)^\mu \cdot \frac{\pi^{2\mu+1}}{(2\mu+1)!} \quad \text{for all } \mu \in I\!N \,,$$

and recognize that $\{A_n\}$ possesses a logarithmic asymptotic expansion of arbitrary order with limit π and $\rho_\mu = 2\mu$ for all μ.

Of course it would make no sense to approximate π by the sequence $\{A_n\}$, if these numbers would have to be computed via (3.3), thus using the number π itself. But fortunately it is possible to calculate certain subsequences of $\{A_n\}$ *recursively.* We shall study this technique in some detail in Sections 3.3 and 5.2, but state here just for completeness the following result:

Define the sequence $\{Y_i\}$ through the recursion formula

$$Y_{i+1} \;:=\; Y_i \cdot \frac{\sqrt{2}}{\sqrt{1+\sqrt{1-(Y_i/2^i)^2}}} \quad \text{for } i = 0,1,2,\ldots, \tag{3.4}$$

with $Y_0 := 2$. Then for all $i \in I\!N$ the identity

$$Y_i \;=\; A_{2^{i+2}}$$

holds; in other words, equation (3.4) yields a recursion formula for the computation of the sequence $\{A_{2^i}\}$.

3.1.2 Continued Fractions of Poincaré Type. Continued fractions are commonly expexted to converge rather quickly; however, this is not true in all cases, as for example A.Hautot [1] illustrated by some examples of "ultra-slowly" converging continued fractions of Poincaré type:

Let us consider the fraction

$$F = 1 + \frac{\alpha_0}{1} + \frac{\alpha_1}{1} + \frac{\alpha_2}{1} + \cdots ,$$

where, for $k > 0$, the coefficients α_k can be written in the form

$$\alpha_k = -\frac{1}{4} + \sum_{\nu=2}^{\infty} \frac{\gamma_\nu}{k^\nu} .$$

With F_n we denote the $n-$ th approximant of F, i.e.

$$F_n = 1 + \frac{\alpha_0}{1} + \frac{\alpha_1}{1} + \cdots + \frac{\alpha_n}{1} .$$

Then the asymptotic structure of the *inverses* of the approximation errors can be characterized as follows:

1. If $1 + 16\,\gamma_2 > 0$, then

$$\frac{1}{F_n - F} = \lambda_0 + \sum_{\nu=1} \frac{\lambda_\nu}{n^{\nu-\tau-1}} , \tag{3.5}$$

with certain coefficients λ_ν and with the constant $\tau = \sqrt{1 + 16\,\gamma_2}$.

But with standard arguments we can show that the sequence $\{F_n\}$ itself possesses a logarithmic asymptotic expansion of the type we are interested in; we obtain that

$$F_n = F + \sum_{\mu=1} \frac{c_\mu}{n^{\rho_\mu}}$$

with $\rho_1 = \tau$, and the other ρ_μ 's being integer multiples of powers of τ .

In the very special case that τ itself is a positive integer, the asymptotic expansion reduces to

$$F_n = F + \sum_{\mu=1} \frac{c_\mu}{n^{\tau+\mu-1}} ;$$

this was already stated in the original paper by A.HAUTOT [1].

2. If $1+16\gamma_2=0$, then the convergence of $\{F_n\}$ is even worse. A. Hautot obtained the asymptotic relation

$$\frac{1}{F_n-F} = \sum_{j=1} \frac{\mu_j}{(\log(n))^j} + \sum_{\nu=0} \frac{\lambda_\nu}{n^\nu} . \tag{3.6}$$

Also here we are able to rewrite this into the form

$$F_n = F + \frac{1}{\log(n)} + \sum_{\substack{j=1\\ \nu=0}} \frac{c_{j\nu}}{(\log(n))^j n^\nu} .$$

Possibly after rearranging the double series on the right hand side, this is also an asymptotic expansion in the sense of Definition 1.3, but it is *not* a logarithmic one!

In a recent paper, A.MATOS [1] developed extrapolation algorithms which exhibit the asymptotic structure of (3.5) resp. (3.6), and therefore can be applied directly to the sequence of inverses $\{(F_n-F)^{-1}\}$. Moreover, S.PASZKOWSKI [1] improved Hautot's method by applying several sequence transformation methods to the continued fraction's tails. More details about this and also the results of some numerical tests can be found in the original paper S.PASZKOWSKI [1].

3.1.3 Numerical Differentiation. Let g denote some complex function, which is analytic in a neighborhood of the point $z \in \mathbb{C}$. Then the sequence of *forward differences*

$$\delta_n^f(z) := n \cdot \big(g(z+1/n) - g(z)\big)$$

is well-defined for n large enough, and possesses the logarithmic asymptotic expansion

$$\delta_n^f(z) = g'(z) + \sum_{\mu=1} \frac{c_\mu(z)}{n^\mu} \tag{3.7}$$

with limit $g'(z)$ and

$$c_\mu(z) \;=\; \frac{g^{(\mu+1)}(z)}{(\mu+1)!} \qquad \text{for all } \mu \in I\!N \;.$$

The validity of (3.7) can easily be seen using the series expansion of g around z.

Quite obviously, this approach allows a lot of modifications and generalizations, which we shall study in Section 3.5. For example, one could replace the forward difference δ_n^f by the *central difference*

$$\delta_n^c(z) \;:=\; \frac{n}{2} \cdot \left(g(z+1/n) - g(z-1/n)\right) \,,$$

which leads to the logarithmic asymptotic expansion

$$\delta_n^c(z) \;=\; g'(z) \;+\; \sum_{\mu=1} \frac{\tilde{c}_\mu(z)}{n^{2\mu}}$$

with limit $g'(z)$ and

$$\tilde{c}_\mu(z) \;=\; \frac{g^{(2\mu+1)}(z)}{(2\mu+1)!} \qquad \text{for all } \mu \in I\!N \;;$$

this idea is due to H.RUTISHAUSER [1].

3.1.4 Asymptotic Expansions for the Error of Best Uniform Polynomial Approximation. In classical approximation theory, there was always great interest in the asymptotic behavior of the error of the best approximation, the so-called *minimal deviation*. We illustrate by two examples that also in this context logarithmic asymptotic expansions play an important role.

Let us restrict ourselves to the interval $[-1,1]$, and let f denote some function from the space $C[-1,1]$ of real continuous functions on this interval; then the minimal deviation, in which we are interested, is defined as the number

$$E_n(f) \;:=\; \inf_{p\in\Pi_n} \{\|f-p\|\} \,, \tag{3.8}$$

where $\|\cdot\|$ denotes the uniform norm on $C[-1,1]$ and Π_n is the space of real polynomials of degree not exceeding n.

Our first example is concerned with a certain class of functions f, containing, for example, the exponential function.

Theorem 3.2 (H.FIEDLER [1]): *Let P denote some power series, which converges near the origin and satisfies $P(0)=1$. Furthermore, let $f(x) = \sum_{\nu=0}^{\infty} \gamma_\nu x^\nu$ be given, such that for all ν large enough*

$$\frac{\gamma_{\nu-1}}{\gamma_\nu} = c \cdot \nu \cdot P(1/\nu) \tag{3.9}$$

with some real constant $c \neq 0$.
Then the sequence $\{2^{n-1} n!\, E_{n-1}(f)\}$ possesses the logarithmic asymptotic expansion

$$2^{n-1} n!\, E_{n-1}(f) = |f^{(n)}(0)| + \sum_{\mu=1} \frac{c_\mu}{n^\mu} . \tag{3.10}$$

The *proof* of Theorem 3.2 is quite long and difficult and would go beyond the scope of this introductory subsection; we refer to the original paper by H.FIEDLER [1].

Example. If

$$f(x) = \mathrm{e}^{ax} \qquad \text{with } a \neq 0 ,$$

then (3.9) is satisfied for all $\nu \in I\!N$ by choosing $P \equiv 1$ and $c = 1/a$. Thus Theorem 3.2 can be applied, and (3.10) yields the existence of the logarithmic asymptotic expansion

$$2^{n-1} n!\, E_{n-1}(\mathrm{e}^{ax}) = a^n + \sum_{\mu=1} \frac{c_\mu}{n^\mu} .$$

This sharpens some older results on the asymptotic behavior of $E_n(\mathrm{e}^x)$, due to G.MEINARDUS [1],

$$2^{n-1} n!\, E_{n-1}(\mathrm{e}^x) = 1 + O(n^{-1}) \quad \text{for } n \to \infty ,$$

and to H.BRASS [1],

$$2^{n-1}\, n!\, E_{n-1}(\mathrm{e}^x) \;=\; 1 + \frac{1}{4n} + O(n^{-2}) \quad \text{for } n \to \infty\,.$$

Let us now turn to another function, whose minimal deviation also always stood in the center of interest in approximation theory, namely $f(x) = |x|$. Since this is an even continuous function, it suffices to consider the values $E_{2n}(|x|)$ only, because

$$E_{2n}(|x|) \;=\; E_{2n+1}(|x|) \quad \text{for all } n \in I\!N_0\,.$$

In 1914, the Russian mathematician S. N. Bernstein proved the existence of a non-zero constant β, such that

$$\lim_{n\to\infty} 2n\, E_{2n}(|x|) \;=\; \beta\,, \tag{3.11}$$

see S.N.BERNSTEIN [1], and deduced furthermore the following rigorous bounds for β:

$$0.278 \;<\; \beta \;<\; 0.286\,.$$

It was already noted by Bernstein himself that the number

$$\frac{1}{2\sqrt{\pi}} \;=\; 0.282\,094\ldots$$

is very close to the arithmetic mean of these two bounds; for this reason, people tried over more than 70 years (in vain) to prove or disprove the so-called

Bernstein conjecture:

$$\beta \;\overset{\text{conj.}}{=}\; \frac{1}{2\sqrt{\pi}}\,. \tag{3.12}$$

Finally, in 1985 R.S.VARGA & A.J.CARPENTER [1] computed the numerical value of the constant β up to 50 decimal places, using very ingenious methods, and showed that Conjecture (3.12) is *false.* (We note in passing that there still does not exist a closed-form expression of β in terms of classical mathematical functions or constants.)

Moreover, inspired by their numerical tests, they stated the following new conjecture.

Conjecture (R.S.VARGA & A.J.CARPENTER [1]): *The sequence* $\{E_{2n}(|x|)\}$ *possesses a logarithmic asymptotic expansion of the form*

$$2n\,E_{2n}(|x|) \;=\; \beta \;+\; \sum_{\mu=1} \frac{c_\mu}{n^{2\mu}} \tag{3.13}$$

with

$$\mathrm{sign}(c_\mu) \;=\; (-1)^\mu \quad \textit{for all} \;\; \mu \in \mathbb{N} \;.$$

In addition, as pointed out by G.MEINARDUS [5], the numerical results indicate the asymptotic relation

$$(-1)^\mu\, c_\mu \;\sim\; (\mu-1)! \;\; \text{for} \quad \mu \to \infty \;. \tag{3.14}$$

To the best of our knowledge, conjectures (3.13) and (3.14) are not yet proved at the moment.

The reader should now have a first impression of logarithmic asymptotic expansions and their appearance in Numerical Analysis. Of course, we could make this list of examples almost arbitrarily long; e.g., think of the recent publications by P.VERLINDEN & A.HAEGEMANS [1] and W.SWELDENS & R.PIESSENS [1,2] on expansions for wavelets. But let us now turn to a more systematic study of logarithmic asymptotic expansions.

3.2 Euler's Summation Formula and Riemann's Zeta-Function

In the present section we have two main goals: First, we shall prove the famous Euler summation formula (sometimes also denoted as Euler-MacLaurin formula), which will be a very important tool at several places in this book, e.g. for the derivation of the asymptotic expansion for the trapezoidal rule (Section 3.4). Then, as another application of the Euler formula, we prove that the partial sums of the infinite series $\zeta(s) = \sum_{\nu=1}^{\infty} \nu^{-s}$, $\mathrm{Re}(s) > 1$, possess a logarithmic asymptotic expansion with limit $\zeta(s)$. Finally, we shall generalize this result to other infinite series

3.2.1 Bernoulli Polynomials. The polynomials we are going to define now play an important role in many different branches of mathematical analysis, cf. for example N.E.NÖRLUND [1]. They were studied the first time by JAK.BERNOULLI [1] at the beginning of the 18th century. In the sequel, we shall list only those properties of them, which will be needed for the proof of Euler's formula; a nice compendium of more results on the Bernoulli polynomials can be found in Nörlund's book and in many textbooks on Numerical Analysis.

Definition 3.3: Define, for all $n \in \mathbb{N}_0$, the polynomials B_n as follows:

$$B_0(x) := 1\,, \tag{3.15}$$

$$B_n(x) := x^n - \frac{1}{n+1} \sum_{\nu=0}^{n-1} \binom{n+1}{\nu} B_\nu(x) \quad \text{for} \quad n = 1, 2, .. \tag{3.16}$$

Then $B_n(x)$ is a polynomial of exact degree n with leading coefficient 1 . It is denoted as n^{th} *Bernoulli polynomial.*

For convenience, we list the first few Bernoulli polynomials :

$$\begin{aligned} B_0(x) &= 1 \\ B_1(x) &= x - \frac{1}{2} \\ B_2(x) &= x^2 - x + \frac{1}{6} \\ B_3(x) &= x^3 - \frac{3}{2}x^2 + \frac{1}{2}x \end{aligned}$$

It is common use to define the Bernoulli polynomials only for $x \in [0,1]$, and we shall also restrict ourselves later to this interval. However, until now our definition as well as the next theorem is valid for all $x \in \mathbb{R}$.

Theorem 3.4: *For all $n \in \mathbb{N}_0$, the Bernoulli polynomials have the following properties:*
a) They satisfy the difference equation

$$B_n(x+1) - B_n(x) \;=\; n\,x^{n-1} . \tag{3.17}$$

b) They satisfy the differential equation

$$(n+1)\,B_n(x) \;=\; B'_{n+1}(x) . \tag{3.18}$$

c) For $n \geq 1$, they are normalized by

$$\int_0^1 B_n(x)\,dx \;=\; 0 . \tag{3.19}$$

Proof. The proofs of all three assertions are straightforward, using induction with respect to n (but note that (3.19) is not true for $n = 0$!). We shall prove only the first one.

Obviously, B_0 and B_1 satisfy the difference equation (3.17), so let $n \geq 2$. Then we obtain from (3.16), using the induction hypothesis:

$$\begin{aligned} B_n(x+1) \;&=\; (x+1)^n \;-\; \frac{1}{n+1}\sum_{\nu=0}^{n-1}\binom{n+1}{\nu}\left(B_\nu(x) + \nu x^{\nu-1}\right) \\ &=\; x^n \;-\; \frac{1}{n+1}\sum_{\nu=0}^{n-1}\binom{n+1}{\nu}B_\nu(x) \\ &\quad + \sum_{\nu=0}^{n-1}\binom{n}{\nu}x^\nu \;-\; \sum_{\nu=0}^{n-1}\frac{\nu}{n+1}\binom{n+1}{\nu}x^{\nu-1} \end{aligned}$$

$$= B_n(x) + n\,x^{n-1}$$
$$+ \sum_{\nu=0}^{n-2} \binom{n}{\nu} x^\nu - \sum_{\nu=1}^{n-1} \frac{\nu}{n+1} \binom{n+1}{\nu} x^{\nu-1}$$
$$= B_n(x) + n\,x^{n-1} .$$

This completes the proof of (3.17) □

It can be shown that the difference equation (3.17) as well as the differential equation (3.18) – the latter one in combination with (3.15) – also *define* a polynomial D_n from Π_n , which is unique except for an additive constant. If, in addition, D_n is normalized such that

$$\int_0^1 D_n(x)\,dx = 0 \quad \text{for } n \geq 1 ,$$

then $D_n = B_n$.

In Theorem 3.4 we have learned about some fundamental properties of the Bernoulli polynomials, in particular that they satisfy a differential as well as a difference equation. In fact, they possess a lot more interesting properties, two of which we present in the next corollary. From now on, we restrict ourselves to the interval $[0, 1]$.

Corollary 3.5: *The Bernoulli polynomials are symmetric with respect to the point* $1/2$. *More precisely we have*

a) For all $x \in [0, 1]$ *and all* $n \in I\!N_0$

$$B_n(1-x) = (-1)^n B_n(x) . \tag{3.20}$$

b) For all $n \in I\!N_0$, $n \neq 1$

$$B_n(0) = B_n(1) . \tag{3.21}$$

Proof. We prove the first assertion by induction with respect to n. For $n = 0$ it is obviously true, so let us assume that (3.20) holds for some $n \in \mathbb{N}_0$. Then also

$$\int_0^t B_n(1-x)\,dx \;=\; (-1)^n \int_0^t B_n(x)\,dx$$

for $t \in [0,1]$, and from (3.18) it follows that

$$-\int_0^t \frac{B'_{n+1}(1-x)}{n+1}\,dx \;=\; (-1)^n \int_0^t \frac{B'_{n+1}(x)}{n+1}\,dx\,.$$

Hence

$$-B_{n+1}(1-t) + B_{n+1}(1) \;=\; (-1)^n \cdot \big(B_{n+1}(t) - B_{n+1}(0)\big)\,,$$

or, in a slightly different form

$$B_{n+1}(1-t) \;=\; (-1)^{n+1} \cdot B_{n+1}(t) + \big(B_{n+1}(1) + (-1)^n B_{n+1}(0)\big)\,. \tag{3.22}$$

Now, integrating both sides of (3.22) between 0 and 1 and using (3.19) gives

$$B_{n+1}(1) + (-1)^n B_{n+1}(0) \;=\; 0\,,$$

and after re-inserting this into (3.22) we have proved the first statement of Corollary 3.5; the second one follows directly from the difference equation (3.17), putting $x = 0$. □

As a direct application of this corollary we obtain the last property of the polynomials B_n we shall make use of, namely

$$B_{2j+1}(0) \;=\; B_{2j+1}(1/2) \;=\; B_{2j+1}(1) \;=\; 0 \text{ for all } j\,. \tag{3.23}$$

Moreover, using induction it is easy to show that these are the only zeros of B_{2j+1} in $[0,1]$.

The values of the Bernoulli polynomials at the origin play an important role; they therefore have a special name:

Definition 3.6: For each $n \in I\!N_0$, the number

$$\beta_n := B_n(0)$$

is denoted as the nth *Bernoulli number.*

It follows that the Bernoulli numbers satisfy the recurrence relation

$$\begin{aligned} \beta_0 &= 1\,, \\ \beta_n &= -\frac{1}{n+1}\sum_{\nu=0}^{n-1}\binom{n+1}{\nu}\beta_\nu \quad \text{for } n=1,2,\dots\,, \end{aligned}$$

and that

$$\beta_{2j+1} = 0 \qquad \text{for } j \geq 1\,, \tag{3.24}$$

i.e., all β_n's with odd index vanish, except for β_1 . It can also be shown (cf. Problem P 3.3) that all Bernoulli numbers with even index are indeed different from zero and alternate in sign, i.e.

$$\text{sign}(\beta_{2j}) = (-1)^{j+1} \qquad \text{for } j \geq 1\,.$$

3.2.2 Euler's Summation Formula. We are now ready to prove Euler's famous summation formula (L.EULER [1]), which provides a relation between the sum of the values of a certain function in discrete points and the integral over this function. The formula will be the major tool for the proofs of some asymptotic relations in the subsequent sections.

Remark. In many approaches and publications, Relation (3.25) is denoted as Euler-McLaurin formula or Euler-McLaurin expansion; to bring in McLaurin's name into this context is justified inasmuch as he found this formula probably without knowing Euler's work. However, since L. Euler was the first one who wrote down this relation, we would like to name it after him alone.

Theorem 3.7 (Euler's summation formula): *Let m and r be nonnegative integers with $m \geq 1$, and let g denote some real function belonging to the class $C^{2r+1}[0,m]$.*
Then the following relation holds:

$$\frac{1}{2}g(0) + \sum_{\nu=1}^{m-1} g(\nu) + \frac{1}{2}g(m) =$$

$$= \int_0^m g(t)\,dt + \sum_{\mu=1}^{r} \frac{\beta_{2\mu}}{(2\mu)!}\left(g^{(2\mu-1)}(m) - g^{(2\mu-1)}(0)\right) \quad (3.25)$$

$$+ \frac{1}{(2r+1)!}\int_0^m B_{2r+1}(t-[t])\cdot g^{(2r+1)}(t)\,dt\,,$$

where we have used Gauß' symbol $[t] := \max\limits_{a\in I\!N}\{a \leq t\}$. *As usual, empty sums are set equal to zero.*

Proof. We prove the assertion by induction with respect to the *differentiability class* r. First, let $r = 0$, and assume that the variable t belongs to the interval $[\nu-1,\nu)$ for some $\nu \in \{1,\ldots,m\}$. Then we obtain for the first Bernoulli polynomial the expression

$$B_1(t-[t]) = t-[t]-\frac{1}{2} = t-\nu+\frac{1}{2}\,,$$

and integration by parts gives us the relation

$$\int_{\nu-1}^{\nu} B_1(t-[t])\,g'(t)\,dt =$$

$$= \left(t-\nu+\frac{1}{2}\right)g(t)\Big|_{\nu-1}^{\nu} - \int_{\nu-1}^{\nu} g(t)\,dt \quad (3.26)$$

$$= \frac{1}{2}\left(g(\nu)+g(\nu-1)\right) - \int_{\nu-1}^{\nu} g(t)\,dt\,.$$

Summing up both sides of (3.26) for $\nu = 1, \ldots, m$ yields

$$\int_0^m B_1(t-[t])\, g'(t)\, dt \;=\; \frac{1}{2}\, g(0) \;+\; \sum_{\nu=1}^{m-1} g(\nu) \;+\; \frac{1}{2}\, g(m) \;-\; \int_0^m g(t)\, dt \;,$$

which was the assertion in the case $r = 0$.

Now let us assume that Equation (3.25) holds for some arbitrary nonnegative integer r ; we have to show that it is also true then for $r+1$ (i.e. we may assume that $g \in C^{2r+3}[0,m]$).
To do this, we again use integration by parts and obtain, after a transformation from $[\nu-1, \nu)$ to $[0,1)$:

$$\int_{\nu-1}^{\nu} B_{2r+1}(t-[t]) \cdot g^{(2r+1)}(t)\, dt \;=$$

$$= \int_0^1 B_{2r+1}(\tau) \cdot g^{(2r+1)}(\tau+\nu-1)\, d\tau$$

$$\overset{(3.18)}{=} \frac{1}{2r+2}\, B_{2r+2}(\tau)\, g^{(2r+1)}(\tau+\nu-1)\Big|_0^1 - \frac{1}{2r+2} \int_0^1 B_{2r+2}(\tau) \cdot g^{(2r+2)}(\tau+\nu-1)\, d\tau$$

$$\overset{(3.21)}{=} \frac{\beta_{2r+2}}{2r+2} \left(g^{(2r+1)}(\nu) - g^{(2r+1)}(\nu-1)\right) - \frac{1}{2r+2} \int_0^1 B_{2r+2}(\tau) \cdot g^{(2r+2)}(\tau+\nu-1)\, d\tau \;.$$

Now we apply the same process to the integrand in the last line and obtain

$$\int_0^1 B_{2r+1}(\tau) \cdot g^{(2r+1)}(\tau + \nu - 1)\, d\tau \ =$$

$$= \frac{\beta_{2r+2}}{2r+2} \left(g^{(2r+1)}(\nu) - g^{(2r+1)}(\nu - 1) \right)$$

$$+ \frac{1}{(2r+2)(2r+3)} \int_0^1 B_{2r+3}(\tau) \cdot g^{(2r+3)}(\tau + \nu - 1)\, d\tau \ ,$$

since $\beta_{2r+3} = 0$, due to (3.24). Finally, a transformation back to the interval $[\nu - 1, \nu)$ and summation for $\nu = 1, \ldots, m$ yields the equation

$$\int_0^m B_{2r+1}(t - [t]) \cdot g^{(2r+1)}(t)\, dt \ =$$

$$= \frac{\beta_{2r+2}}{2r+2} \left(g^{(2r+1)}(m) - g^{(2r+1)}(0) \right) + \frac{1}{(2r+2)(2r+3)} \int_0^m B_{2r+3}(t - [t]) \cdot g^{(2r+3)}(t)\, dt \ .$$

This shows – after inserting into the induction hypothesis – the validity of (3.25) for $r + 1$ and thus completes our proof. □

In the nice habilitation thesis by C.SCHNEIDER [2], one can find several other forms of the remainder term and the coefficients in Euler's summation formula, which do not involve derivatives of the function g. Instead, the concept of (generalized) divided differences is used, which leads to the development of algorithms for the automatic computation of error bounds, remainder terms and even asymptotic expansions.

3.2.3 Asymptotic Expansion for the Partial Sums of Riemann's Zeta-Function. One of the outstanding higher transcendental functions of mathematical analysis is surely *Riemann's* *zeta-function* $\zeta(s)$ (see B.RIEMANN [1]), which can be defined for all $s \in \mathbb{C}$ with $\mathrm{Re}(s) > 1$ through the convergent infinite series

$$\zeta(s) := \sum_{\nu=1}^{\infty} \frac{1}{\nu^s} . \tag{3.27}$$

There exists a vast literature on this function, and the basics about it can be found in any textbook on complex analysis. It is well-known that $\zeta(s)$ can be continued into the complex plane, and that this continuation – which we also denote by $\zeta(s)$ – is a meromorphic function.

One of the main reasons why this function stimulated for all the years the scientific interest of so many mathematicians is the existence of the famous *Riemann Hypothesis*, which says that the each zero s_0 of ζ in the so-called *critical strip*

$$\{s \in \mathbb{C} \mid 0 \leq \mathrm{Re}(s) \leq 1\}$$

satisfies

$$\mathrm{Re}(s_0) = \frac{1}{2} .$$

This hypothesis is not yet proved nor disproved until now, although many people spent a lot of time and energy on it, and produced in the meantime a whole catalogue of hypotheses, which are all equivalent to Riemann's original one.

In contrast to the huge number of theoretical investigations on ζ stands the number of existing methods for its numerical calculation, and in Part II of this book we shall study a quite efficient one, which is based on the extrapolation technique. In preparation for this (but not only in preparation, since the result is also of some theoretical importance) we shall prove now that the partial sums of the infinite series (3.27) possess a logarithmic asymptotic expansion with limit – of course – $\zeta(s)$.

Theorem 3.8 (cf. G.MEINARDUS [3]): *Let* s *be some complex variable with* $\mathrm{Re}(s) > 1$ *, and denote by* $\zeta_n(s)$ *the* n^{th} *partial sum of the series (3.27), i.e.*

$$\zeta_n(s) := \sum_{\nu=1}^{n} \frac{1}{\nu^s} . \tag{3.28}$$

Then $\{\zeta_n(s)\}$ *possesses the following logarithmic asymptotic expansion of arbitrary order:*

$$\zeta_n(s) = \zeta(s) - \frac{(s-1)^{-1}}{n^{s-1}} + \frac{2^{-1}}{n^s} + \sum_{\mu=1} \frac{c_{2\mu}(s)}{n^{s+2\mu-1}} \tag{3.29}$$

with

$$c_{2\mu}(s) = -\frac{\beta_{2\mu}}{(2\mu)!} s(s+1)\cdots(s+2\mu-2) \quad \textit{for all} \quad \mu \in I\!N .$$

Here $\beta_{2\mu}$ *again denotes the* $(2\mu)^{th}$ *Bernoulli number.*

Proof. With an arbitrary, but for the moment fixed $n \in I\!N$, we consider the real function

$$g(t) := \frac{1}{(n+t)^s} .$$

Obviously $g \in C^r[0,\infty)$ for each $r \in I\!N_0$, and

$$g^{(2\mu-1)}(t) = \frac{-s(s+1)\cdots(s+2\mu-2)}{(n+t)^{s+2\mu-1}}$$

for all $\mu \in I\!N$. Therefore we may apply Euler's summation formula (3.25) (with $m \to \infty$) to this function g and obtain

$$\frac{1}{2n^s} + \sum_{\nu=1}^{\infty} \frac{1}{(n+\nu)^s} = \int_0^{\infty} \frac{1}{(n+t)^s}\, dt \tag{3.30}$$

$$+ \sum_{\mu=1}^{r} \frac{\beta_{2\mu}}{(2\mu)!} \frac{s(s+1)\cdots(s+2\mu-2)}{n^{s+2\mu-1}} + o(n^{-s-2r+1}) .$$

But since the differentiability order r can be chosen arbitrarily large, and due to the facts that

$$\int_0^\infty \frac{1}{(n+t)^s}\, dt = \frac{(s-1)^{-1}}{n^{s-1}} \quad \text{and}$$

$$\sum_{\nu=1}^{\infty} \frac{1}{(n+\nu)^s} = \zeta(s) - \zeta_n(s) \,,$$

Equation (3.30) already yields the assertion of Theorem 3.8. □

From the first term of the expansion (3.29) one can read off very easily some facts on the convergence behaviour of the infinite series (3.27). For example, we see that the partial sums $\zeta_n(s)$ converge like $n^{Re(s)-1}$, i.e. the convergence gets terribly slow if s approaches the critical strip. Moreover, if s tends to the number 1 itself, the also the asymptotic constant gets infinitely large: The series has turned into the *harmonic series*, where convergence is no longer given, and $\zeta(s)$ possesses a first-order pole in this point.

We shall supply a numerical example for Theorem 3.8 in connection with the corresponding extrapolation process in Part II (Subsection 4.2.2).

3.2.4 Generalization to Other Infinite Series. Although there exist a lot of investigations on the asymptotic behavior of infinite series and their partial sums, and in spite of the fact that many convergence acceleration methods for this problem can be found (cf. for example J.E.DRUMMOND [1,2], S.LUBKIN [1], H.E.SALZER [1], D.SHANKS [1], P.WYNN [1], and in particular the monographs by C.BREZINSKI [1] or J.WIMP [1] and the references therein), the existence of an asymptotic expansion for the partial sums of such series is quite rarely proved. But this is a very essential condition for the applicability of extrapolation methods for convergence acceleration.

For a certain type of infinite series we are now in the position to prove the existence of such an asymptotic expansion by using Theorem 3.8 (and generalizing it at the same time).

Theorem 3.9 (G.WALZ [1]): *Consider a sequence of numbers* $\{u_\nu\}$ *such that the infinite series*

$$S := \sum_{\nu=1}^{\infty} u_\nu$$

converges. Then the following two assertions are equivalent:
(i) The partial sums

$$\sigma_n := \sum_{\nu=1}^{n} u_\nu$$

possess a logarithmic asymptotic expansion of the form

$$\sigma_n = S + \sum_{\mu=1} \frac{c_\mu}{n^{\rho_\mu}} . \tag{3.31}$$

(ii) The elements u_n *possess a logarithmic asymptotic expansion of the form*

$$u_n = \sum_{\mu=1} \frac{d_\mu}{n^{\lambda_\mu}} \tag{3.32}$$

with $\mathrm{Re}(\lambda_1) > 1$.

Proof. Assume that (3.31) is valid; then we can write

$$u_n = \sigma_n - \sigma_{n-1} = \sum_{\mu=1} c_\mu \left(\frac{1}{n^{\rho_\mu}} - \frac{1}{(n-1)^{\rho_\mu}} \right) . \tag{3.33}$$

In order to develop the right hand side into an expansion of the desired form, we make use of the binomial function's series expansion (see e.g. K.KNOPP [1]). This yields, for any exponent ρ,

$$\begin{aligned} \frac{1}{(n-1)^\rho} &= \frac{1}{n^\rho} \cdot \frac{1}{(1-1/n)^\rho} \\ &= \frac{1}{n^\rho} \cdot \left(1 + \sum_{\kappa=1}^{\infty} \binom{-\rho}{\kappa} (-1)^\kappa \frac{1}{n^\kappa} \right) := \sum_{\kappa=0} \frac{a_\kappa}{n^{\rho+\kappa}} , \end{aligned}$$

and by inserting this into (3.33) we obtain

$$u_n = \sum_{\mu=1} c_\mu \sum_{\kappa=1} \frac{-a_\kappa}{n^{\rho_\mu+\kappa}} := \sum_{\mu=1} \frac{d_\mu}{n^{\lambda_\mu}}$$

with $\lambda_1 = \rho_1 + 1$. Since (3.31) is a logarithmic asymptotic expansion we know that $\mathrm{Re}(\rho_1) > 0$, hence $\mathrm{Re}(\lambda_1) > 1$.

To deduce (3.31) from (3.32) we analyze the difference $\sigma_n - S$. This gives us

$$\sigma_n - S = -\sum_{\nu=n+1}^{\infty} u_\nu = -\sum_{\nu=n+1}^{\infty} \sum_{\mu=1} \frac{d_\mu}{\nu^{\lambda_\mu}}$$

$$= -\sum_{\mu=1} d_\mu \sum_{\nu=n+1}^{\infty} \frac{1}{\nu^{\lambda_\mu}} . \tag{3.34}$$

Now one recognizes that the last term in (3.34) is nothing else but the tail of the nth partial sum $\zeta_n(\lambda_\mu)$ of the zeta-function $\zeta(\lambda_\mu)$ (note that $\mathrm{Re}(\lambda_\mu) > 1$ for all μ!), and therefore Theorem 3.8 enables us to develop it into an expansion of the form desired:

$$\sum_{\nu=n+1}^{\infty} \frac{1}{\nu^{\lambda_\mu}} = \zeta(\lambda_\mu) - \zeta_n(\lambda_\mu) \overset{(3.29)}{=} \sum_{\kappa=0} \frac{\gamma(\lambda_\mu)}{n^{\lambda_\mu - 1 + \kappa}} .$$

Finally, we replace the last term in (3.34) by this expansion and obtain

$$\sigma_n - S = -\sum_{\mu=1} d_\mu \sum_{\kappa=0} \frac{\gamma(\lambda_\mu)}{n^{\lambda_\mu-1+\kappa}} =: \sum_{\nu=1} \frac{c_\nu}{n^{\rho_\nu}}$$

with $\rho_1 = \lambda_1 - 1$, hence $\mathrm{Re}(\rho_1) > 0$. □

3.3 Asymptotic Expansions with a Prescribed Limit Function

In the preceeding sections we always started with a given sequence $\{\sigma_n\}$ and tried to prove the existence of an asymptotic expansion for it. Now we shall go the other way round, i. e. we start with a prescribed function, say f, and try to *construct* a sequence, which possesses a logarithmic asymptotic expansion with limit f. In order to attack this problem we use two different approaches, the first of which uses infinite product representations of entire functions and relies on a famous theorem by K. Weierstraß.

3.3.1 Partial Products of Infinite Products. Let us begin with the following special case of the Weierstraß factorization theorem:

Theorem 3.10 (K.WEIERSTRASS [1]): *Consider some arbitrary entire function* f *with* $f(0) \neq 0$, *and denote the set of its zeros by* $\{a_\nu\}$. *Furthermore, let there exist some natural number* r, *such that the infinite series*

$$\sum_{\nu=1}^{\infty} \left(\frac{1}{|a_\nu|}\right)^{r+1}$$

converges and set, for all ν *in* $\mathbb{N}$ *:*

$$E_\nu(z) := \left(1 - \frac{z}{a_\nu}\right) \cdot \exp\left(\sum_{j=1}^{r} \frac{1}{j}\left(\frac{z}{a_\nu}\right)^j\right) . \tag{3.35}$$

Then there exists an entire function g, *such that for all* $z \in \mathbb{C}$ *the equality*

$$f(z) = \exp(g(z)) \cdot \prod_{\nu=1}^{\infty} E_\nu(z) \tag{3.36}$$

holds.

Proof. See any textbook on complex analysis, e.g. J.B.CONWAY [1], pp. 170 or R.REMMERT [1], pp. 67. □

Remark. If the function f happens to possess a zero of multiplicity k in $z=0$, then Theorem 3.10 obviously applies to the function

$$\tilde{f}(z) \;:=\; \frac{f(z)}{z^k}\,.$$

In the present form, Theorem 3.10 is only a special case and allows many modifications and generalizations (for example, one can allow the number r to depend on ν, or can restrict to functions which are holomorphic only in some open subset of $\mathbb{C}$), and the same is true for the following constructions.

The connection of Weierstraß' theorem with our considerations on asymptotic expansions is given by the following Theorem 3.11, which says that – under certain assumptions on the set $\{a_\nu\}$ – the sequence of partial products

$$\sigma_n(z) \;:=\; \exp(g(z)) \cdot \prod_{\nu=1}^{n} E_\nu(z) \tag{3.37}$$

possesses a logarithmic asymptotic expansion with limit $f(z)$.

Theorem 3.11 (G.WALZ [1]): *Adopt the notations and assumptions of Theorem 3.10, and assume in addition that for all $j \geq r+1$ the partial sums of the series $\sum_{\nu=1}^{\infty} \left(\frac{1}{a_\nu}\right)^j$ possess a logarithmic asymptotic expansion, i.e. there exist numbers $\eta(j)$ and $c_\mu(j)$, such that*

$$\sum_{\nu=1}^{n} \left(\frac{1}{a_\nu}\right)^j \;=\; \eta(j) \;+\; \sum_{\mu=1} \frac{c_\mu(j)}{n^{\rho_\mu(j)}}\,. \tag{3.38}$$

Then also the partial products in (3.37) possess an asymptotic expansion of the form

$$\sigma_n(z) \;=\; f(z) \;+\; \sum_{\mu=1} \frac{\gamma_\mu(z)}{n^{\lambda_\mu}}\,, \tag{3.39}$$

where the λ_μ *are sums of certain* $\rho_\mu(j)'s$ *, and in particular* $\lambda_1 = \rho_1(r+1)$.

Proof. We may assume $\exp(g(z)) \equiv 1$. Then the application of the natural logarithm to $\sigma_n(z)$ yields

$$
\begin{aligned}
\log(\sigma_n(z)) &= \log\left(\prod_{\nu=1}^{n} E_\nu(z)\right) \\
&= \sum_{\nu=1}^{n}\left(\log\left(1-\frac{z}{a_\nu}\right)+\sum_{j=1}^{r}\frac{1}{j}\left(\frac{z}{a_\nu}\right)^j\right) \\
&= \sum_{\nu=1}^{n}\left(-\sum_{j=1}^{\infty}\frac{z^j}{j\cdot a_\nu^j}+\sum_{j=1}^{r}\frac{1}{j}\left(\frac{z}{a_\nu}\right)^j\right) \\
&= -\sum_{j=r+1}^{\infty}\frac{z^j}{j}\sum_{\nu=1}^{n}\frac{1}{a_\nu^j} \\
&= -\sum_{j=r+1}^{\infty}\frac{z^j}{j}\left(\eta(j)+\sum_{\mu=1}\frac{c_\mu(j)}{n^{\rho_\mu(j)}}\right) .
\end{aligned}
$$

Now we go back to the function $\sigma_n(z)$ itself, i.e. we apply the exponential function to the above expressions; this gives us

$$
\begin{aligned}
\sigma_n(z) &= \exp\left(-\sum_{j=r+1}^{\infty}\frac{z^j}{j}\left(\eta(j)+\sum_{\mu=1}\frac{c_\mu(j)}{n^{\rho_\mu(j)}}\right)\right) \\
&= \sum_{l=0}^{\infty}\frac{1}{l!}\cdot\left(-\sum_{j=r+1}^{\infty}\frac{z^j}{j}\left(\eta(j)+\sum_{\mu=1}\frac{c_\mu(j)}{n^{\rho_\mu(j)}}\right)\right)^l \\
&=: f_1(z) + \sum_{\mu=1}\frac{\gamma_\mu(z)}{n^{\lambda_\mu}}
\end{aligned}
$$

with some function f_1, which does not depend on n. Finally, taking on both sides the limits for $n \to \infty$, we obtain $f_1 \equiv f$. $\Box$

Remark. In view of Theorem 3.9, condition (3.38) could as well be replaced by the assumption that the terms $1/a_\nu^j$ possess a logarithmic asymptotic expansion

$$\frac{1}{a_\nu^j} = \sum_{\mu=1} \frac{d_\mu(j)}{n^{\lambda_\mu(j)}}$$

with $\mathrm{Re}(\lambda_1(j)) > 1$ for all j.

We conclude this section giving an *example* for Theorem 3.11. With $\Gamma(z)$ denoting the usual Gamma-function, we consider the entire function

$$f(z) := \frac{1}{z \cdot \Gamma(z)} .$$

It is well-known that the zeros of f are precisely the negative integers, each of them having multiplicity one. Therefore, the assumptions in Weierstraß' factorization theorem are satisfied, and since

$$\sum_{\nu=1}^{\infty} \frac{1}{\nu^2} < \infty ,$$

we may take $r = 1$. Furthermore, the exponential function in front of the Weierstraß product can be chosen to be $\exp(g(z)) := \exp(Cz)$ with the so-called Euler-Mascheroni constant C (cf. Problem P 3.4).
In order to check the applicability of Theorem 3.11, we have to look onto the asymptotic behaviour of the sums $\sum_{\nu=1}^{n} \frac{1}{(-\nu)^j}$ for all $j \in I\!N$, $j \geq 2$. But thanks to Theorem 3.8, we obtain immediately

$$\sum_{\nu=1}^{n} \frac{1}{(-\nu)^j} = (-1)^j \, \zeta(j) + \sum_{\mu=1} \frac{c_\mu(j)}{n^{j-2+\mu}} ,$$

i.e. a relation of the type (3.38). Therefore we may apply Theorem 3.11 to the function f under consideration and obtain the following result: The partial products

$$\sigma_n(z) \;:=\; \exp(Cz) \cdot \prod_{\nu=1}^{n} \left((1 + \frac{z}{\nu}) \, \exp(\frac{-z}{\nu}) \right)$$

of the Weierstraß product for the function f possess the logarithmic asymptotic expansion

$$\sigma_n(z) \;=\; \frac{1}{z\,\Gamma(z)} \;+\; \sum_{\mu=1} \frac{\gamma_\mu(z)}{n^\mu} \,. \tag{3.40}$$

Another example for Theorem 3.11 is investigated in Problem P 3.6.

3.3.2 Use of Taylor's Expansion. Archimedes' method for the numerical computation of π (cf. Subsection 3.1.1) can obviously be modified and generalized in many ways; in the present subsection we use the idea behind it in order to construct asymptotic expansions for real functions, which are invertible on a certain interval and possess there a series expansion.

Most of the material presented in this subsection comes from the author's doctoral thesis (G.WALZ [1], see also G.WALZ [3]), but special cases of it can already be found in the papers of H.RUTISHAUSER [1] and S.FILIPPI [1]. Also A.HURWITZ [1] presented in 1911 an iterative method for the computation of elementary functions, which is strongly connected with our approach, but Hurwitz of course did not ask for the existence of an asymptotic expansion.

We begin with the following illustrative example, which is a direct generalization of Archimedes' method:

Example (G.WALZ [1,3], S.FILIPPI [1]). Our aim is to construct a logarithmic asymptotic expansion with limit

$$f(x) \;=\; \arcsin(x) \qquad \text{for } x \in [-1,1] \,.$$

In this special case we know explicitly the series expansion for the *inverse* $f^{-1}(y) = \sin(y)$ of f, namely

$$f^{-1}(y) \;=\; y \;+\; \sum_{\nu=1}^{\infty} \frac{(-1)^{\nu}}{(2\nu+1)!} \cdot y^{2\nu+1} \tag{3.41}$$

for all $y \in I\!R$. So, if we define, for $n \in I\!N$ and $x \in [-1,1]$

$$\sigma_n(x) \;:=\; n \cdot \sin\left(\frac{\arcsin(x)}{n}\right) , \tag{3.42}$$

we immediately get from (3.41) the existence of an asymptotic expansion of the form

$$\sigma_n(x) \;=\; \arcsin(x) \;+\; \sum_{\mu=1} \frac{c_{\mu}(x)}{n^{2\mu}} , \tag{3.43}$$

i.e. a logarithmic asymptotic expansion of arbitrary order and with the desired limit.

At first sight, this result seems to be only of theoretical importance, because if one would approximate the arcsin-function with the sequence $\{\sigma_n\}$ as defined in (3.42), he would have to evaluate this function and its inverse in each step. But fortunately it is possible, very much like in Subsection 3.1.1, to compute the subsequence $\{\sigma_{2^i}(x)\}$ for all $i \in I\!N_0$ *recursively*:

Lemma 3.12: *For arbitrary $x \in [-1,1]$, define the sequence $\{y_i\}$ through*

$$y_0 \;:=\; x \,, \qquad \text{and for } i = 1,2,\ldots$$

$$y_i \;:=\; \frac{\sqrt{2}\cdot y_{i-1}}{\sqrt{1+\sqrt{1-(y_{i-1}/2^{i-1})^2}}} \,. \tag{3.44}$$

Then for all $i \in I\!N_0$:

$$y_i \;=\; y_i(x) \;=\; \sigma_{2^i}(x)$$

with σ defined in (3.42).

Proof. Use (3.42) and the half-angle formula of the sine function (see e.g. M. ABRAMOWITZ & I. STEGUN [1]). □

One easily recognizes the general principle standing behind this example as well as behind the construction in Subsection 3.1.1.; this motivates our next theorem, which is the central one of the present subsection.

Theorem 3.13 (G. Walz [1,3]): *Let there be given a real interval I, of which the origin is an interior point, and consider some real function f, which maps I bijectively onto an interval J. Furthermore, assume that f possesses for all $x \in I$ the series expansion*

$$f(x) \;=\; \sum_{\nu=1}^{\infty} a_\nu x^\nu$$

with $a_1 \neq 0$. Then the sequence of functions $\{\sigma_n(x)\}$, defined for all $n \in \mathbb{N}$ through

$$\sigma_n(x) \;:=\; n \cdot a_1 \cdot f^{-1}\left(\frac{f(x)}{n}\right), \tag{3.45}$$

possesses a logarithmic asymptotic expansion of arbitrary order with limit $f(x)$.

Proof. Since f is one-to-one, there exists a function f^{-1} on J, which has for all $y \in J$ a series expansion of the form

$$f^{-1}(y) \;=\; \sum_{j=1}^{\infty} b_j y^j \tag{3.46}$$

(note that $0 \in J$, since $f(0) = 0$). From the identity

$$x \;=\; \sum_{j=1}^{\infty} b_j \left(\sum_{\nu=1}^{\infty} a_\nu x^\nu\right)^j$$

we see that the coefficients b_j are rational functions of the $a'_\nu s$, e.g.

$$b_1 = \frac{1}{a_1} \quad \text{and} \quad b_2 = -\frac{a_2}{a_1^3} .$$

Therefore, using (3.46), it follows directly from the definition of σ_n that

$$\sigma_n(x) = n \cdot a_1 \cdot f^{-1}\left(\frac{f(x)}{n}\right) = \frac{n}{b_1} \cdot \sum_{j=1}^{\infty} b_j \left(\frac{f(x)}{n}\right)^j$$

$$= f(x) + \sum_{j=2}^{\infty} b_j \frac{f(x)^j}{n^{j-1}} =: f(x) + \sum_{\mu=1} \frac{c_\mu(x)}{n^\mu}, \quad (3.47)$$

which is of the desired form. □

Obviously, Theorem 3.13 allows a lot of modifications; we will indicate a few of these in the following remarks.

1. The structure of the series expansion of f^{-1} carries over very closely to that of the resulting asymptotic expansion. If, for example, the series of f^{-1} contains only odd powers of y, then the asymptotic expansion (3.47) contains only even powers of n (cf. the arcsin-example from above), a very desirable situation in extrapolation theory.
However, also in the general case it is possible to produce asymptotic expansions of this type, namely by defining

$$\tilde{\sigma}_n(x) := \frac{n \cdot a_1}{2} \cdot \left(f^{-1}\left(\frac{f(x)}{n}\right) - f^{-1}\left(\frac{f(x)}{-n}\right)\right) . \quad (3.48)$$

In Subsection 3.5.3 we shall come back to constructions of this type in more detail.
2. If the functions f resp. f^{-1} do not allow a whole Taylor expansion, but only a Taylor polynomial of degree, say, $m+1$, then the resulting logarithmic asymptotic expansion is of the finite order m (Problem P 3.7).

3. The preceeding considerations, including Theorem 3.13, can be stated for the complex case as well; the necessary results for complex series expansions and their inverses can be found in many textbooks, e.g. G.SANSONE & J.GERETSEN [1].

A closer look into the proof of Theorem 3.13 shows that one could as well replace the function $f(x)$ in (3.45) by any other function $g(x)$ (and a_1 by $1/b_1$), and would still obtain a logarithmic asymptotic expansion, in this case of course with limit $g(x)$. For some special cases this approach was proposed by S.FILIPPI [1] and H.RUTISHAUSER [1].

However, this method has in general a serious drawback: As we shall see, the elements of the subsequence $\{\sigma_{2^i}(x)\}$ can in many cases be computed recursively, provided that the first one, $\sigma_1(x)$, is known. Now, with our approach this is easy, since always $\sigma_1(x) = a_1\, x$, but in the other case one only has that

$$\sigma_1(x) \;=\; a_1\, f^{-1}(g(x)) \;,$$

which can only in very special cases be computed excplicitly.

Let us continue with another *example* for Theorem 3.13. The function

$$f(x) \;:=\; \log(1+x) \;=\; \sum_{j=1}^{\infty} \frac{(-1)^{j+1}}{j}\, x^j$$

satisfies the assumptions of the theorem on any interval $I = [-\alpha, \alpha]$ with $0 < \alpha < 1$, its inverse function being

$$f^{-1}(y) \;=\; \exp(y) - 1 \;.$$

Therefore the functions

$$\sigma_n(x) \;:=\; n \cdot \left(\exp\left(\frac{\log(1+x)}{n} \right) - 1 \right)$$

possess the logarithmic asymptotic expansion

$$\sigma_n(x) \;=\; \log(1+x) \;+\; \sum_{\mu=1} \frac{c_\mu(x)}{n^\mu} \;.$$

We remark that in this case *all* elements $\sigma_n(x)$ can be computed by elementary operations, since for all $n \in \mathbb{N}$ we have

$$\sigma_n(x) \; = \; n \cdot ((1+x)^{1/n} - 1) \, ,$$

a well-known formula for the numerical computation of the natural logarithm.
Finally, an application of the construction indicated in (3.48) leads us to the functions

$$\tilde{\sigma}_n(x) \; = \; \frac{n}{2} \cdot \left((1+x)^{1/n} - (1+x)^{-1/n} \right) \, ,$$

which possess an expansion of the type

$$\tilde{\sigma}_n(x) \; = \; \log(1+x) \; + \; \sum_{\mu=1} \frac{c_{2\mu}(x)}{n^{2\mu}} \; .$$

Many interesting scalar functions can be re-defined as functions of matrices, and there exist different natural ways to do this. One of them is given in the following theorem, to be found, for example, in E.STICKEL [1]:

Theorem 3.14: *For a complex number* ζ *and a real number* $\rho > 0$ *let* $K_\rho(\zeta)$ *denote the set* $\{z \in \mathbb{C};\, |z-\zeta| < \rho\}$ *. Consider a scalar function* f *, which possesses for all* $z \in K_\rho(\zeta)$ *a convergent series expansion of the form*

$$f(z) \; = \; \sum_{\nu=0}^{\infty} \alpha_\nu \, (z-\zeta)^\nu \; .$$

Then for each quadratic matrix A *, whose eigenvalues all lie in* $K_\rho(\zeta)$ *, the matrix function*

$$f(A) \; = \; \sum_{\nu=0}^{\infty} \alpha_\nu \, (A - \zeta I)^\nu \qquad (3.49)$$

is well-defined.

Therefore functions such as $\exp(A)$, $\cos(A)$ and $\sin(A)$ are well-defined for all quadratic matrices A . The *numerical calculation* of a general matrix function is a problem of itself, see, for example, the studies of C.MOLER & C.VAN LOAN [1] or E.STICKEL [1]. In particular, the partial sums of (3.49) are not suitable in most cases; a much better way is in many cases the use of an extrapolation method (cf. G.WALZ [2] or G.F.THOMAS [1]). The crucial point for the applicability of such a method is the existence of a sequence of matrix functions $\{\sigma_n(A)\}$ with an asymptotic expansion. Here the following theorem may be helpful.

Theorem 3.15 (G.WALZ [2]): *Let $f(z)$ be a scalar function, which is one-to-one on some set $K_\rho(0)$, having there a convergent series expansion of the form*

$$f(z) = \sum_{\nu=1}^{\infty} a_\nu z^\nu$$

with $a_1 \neq 0$.
Then for each matrix A, whose eigenvalues all lie in $K_\rho(0)$, the function $f(A)$ and the sequence of functions

$$\sigma_n(A) := n \cdot a_1 \cdot f^{-1}\left(\frac{1}{n} \cdot f(A)\right)$$

are well-defined. Furthermore, the sequence $\{\sigma_n(A)\}$ possesses the logarithmic asymptotic expansion

$$\sigma_n(A) = f(A) + \sum_{\mu=1} \frac{1}{n^\mu} c_\mu(A)$$

with matrix functions $c_\mu(A)$.

Another very important matrix function is the *matrix exponential,* which appears for example in the context of differential equation systems. We shall see in Subsection 3.6.1 that also for this matrix function an approximating sequence with a logarithmic asymptotic expansion can be constructed. Moreover, an extrapolation algorithm for the numerical computation of the matrix exponential is given in Subsection 4.3.3.

There do exists a lot more examples and applications for Theorem 3.13, concerning all kinds of elementary functions, such as area-functions or $\exp(x)$. But let us turn now to a certain interesting class of higher transcendental functions:

3.3.3 Elliptic Functions and Elliptic Integrals. It was already indicated several times that the approximation of the function f via asymptotic expansions as defined in (3.45) is very efficient, if the subsequence $\{\sigma_{2^i}\}$ can be computed recursively. This leads us directly to the question, for which functions f the inverse function f^{-1} satisfies an *algebraic addition theorem.* The answer to this question was given already a hundred years ago by

K.WEIERSTRASS [2], who, roughly spoken, proved that the biggest class of functions, which satisfy an algebraic addition theorem, is that of the doubly-periodic or elliptic functions; we will now give a precise definition of this term.

Definition 3.16: A meromorphic function $\varphi : \mathbb{C} \to \mathbb{C}^*$ is called *elliptic function*, if there exist two non-zero complex numbers ω_1, ω_2 with $\omega_1/\omega_2 \notin \mathbb{R}$, such that for all $z \in \mathbb{C}$

$$\varphi(z) = \varphi(z+\omega_1) = \varphi(z+\omega_2) .$$

If the constants ω_1 and ω_2 are the smallest ones with this property, they are denoted as *fundamental periods* of φ. Furthermore, for any fixed complex number a, the set

$$P_a(\varphi) := \{z \in \mathbb{C};\ z = a + \lambda_1\omega_1 + \lambda_2\omega_2;\ 0 \leq \lambda_1, \lambda_2 < 1\}$$

is denoted as *(fundamental) period-parallelogram* of φ with respect to a, or simply as a *cell* of φ.

In the following we sketch only those properties of elliptic functions which will be needed for the ongoing considerations. The reader interested in details on these functions or in the proofs of the following assertions is kindly referred to the literature on this topic, e.g. the monographs A.CAYLEY [1], K.CHANDRASEKHARAN [1] or E.WHITTAKER & G.WATSON [1].

It is an immediate consequence of Liouville's theorem that an elliptic function φ without poles can only be a constant. Furthermore, it can also be shown that the sum of the residues of φ in each parallelogram $P_a(\varphi)$ equals zero. This implies that every non-constant elliptic function φ must have at least two poles (counted according to their multiplicities) in each cell $P_a(\varphi)$. Therefore, there are two ways to define a "simple" elliptic function, at least as far as singularities are concerned. The first choice is to take functions with one irreducible double pole, at which the residue is zero; this approach is usually named after Weierstraß himself.

The other possibility, onto which we will concentrate now, is to take functions with two simple poles in each cell, at which the residues are of the same modulus but opposite in sign. Elliptic functions of this type are called *Jacobian elliptic functions.* In the following we shall study one of these functions, the *sinus amplitudinis ("sine of the amplitude")*, in some detail; it plays a fundamental role in the theory of Jacobian elliptic functions, very much like the sine function does within the theory of trigonometric functions.

Consider, for some real numbers α and k, $0 \leq k \leq 1$, the integral

$$y := \int_0^\alpha \frac{d\vartheta}{\sqrt{1-k^2 \cdot \sin^2 \vartheta}} . \tag{3.50}$$

Then the function $\alpha(y,k)$, which is implicitly defined by the integral (3.50), is denoted as the *amplitude* of y, and the sinus amplitudinis $\operatorname{sn}(y,k)$ is just the ordinary sine of this amplitude, i.e.

$$\operatorname{sn}(y,k) := \sin(\alpha(y,k)) .$$

We have $\operatorname{sn}(y,0) = \sin(y)$ and $\operatorname{sn}(y,1) = \tanh(y)$. The function sn therefore generalizes the trigonometric as well as the hyperbolic functions in a quite natural way.

So far our definition of $\operatorname{sn}(y,k)$ is valid only for real values of y, and this is the case we shall concentrate on; but this function can without difficulties be continued into the whole complex plane. This continuation process leads to an elliptic function, which we shall also denote by sn. Introducing the *complete elliptic integral of the first kind*

$$K(k) := \int_0^1 \frac{dt}{\sqrt{(1-t^2)(1-k^2t^2)}}$$

and the notation $k' := \sqrt{1-k^2}$, one can show that the function $\operatorname{sn}(z,k)$ has the fundamental periods $\omega_1 = 4\,K(k)$ and $\omega_2 = 2\,i\,K(k')$ (which means that the period parallelogram in this cases is a rectangle), and possesses first-order poles precisely in the points

$$z_{\nu,\mu} \;:=\; 2\nu\,K(k) \;+\; (2\mu+1)\,i\,K(k'), \qquad \nu,\mu \in \mathbb{Z}\ .$$

Since it is an elliptic function, sn satisfies an algebraic addition theorem (see e.g. M. ABRAMOWITZ & I. STEGUN [1]), from which we deduce the following "half-angle-formula" (whatever an angle in this context may be):

$$\operatorname{sn}^2\Big(\frac{y}{2},k\Big) \;=\; \frac{1-\sqrt{1-\operatorname{sn}^2(y,k)}}{1+\sqrt{1-k^2\operatorname{sn}^2(y,k)}}\ . \tag{3.51}$$

In order to apply the results of the preceeding subsection to the function sn , we still need some information about its inverse function; but it is a standard result from the theory of ellliptic functions that this inverse is is just the *elliptic integral of the first kind*, defined through

$$F(x,k) \;:=\; \int_0^x \frac{dt}{\sqrt{(1-t^2)(1-k^2t^2)}}$$

(note that $F(1,k) = K(k)$), a function which is also of great interest in many disciplines. In addition, it is well-known that $F(x,k)$ has a series expansion of the form

$$F(x,k) \;=\; x \;+\; \sum_{\nu=1}^{\infty} a_\nu(k)\,x^{2\nu+1}\ . \tag{3.52}$$

We are now in the position to summarize our considerations in the following theorem, which yields an asymptotic expansion with limit $F(x,k)$. It should be regarded as an instructive example for the application of our results to general elliptic functions and integrals.

Theorem 3.17 (G.Walz [1,3]): *The sequence of functions* $\{\sigma_n(x,k)\}$, *defined for all* $n \in \mathbb{N}$ *through*

$$\sigma_n(x,k) \;:=\; n \cdot \mathrm{sn}\left(\frac{F(x,k)}{n}, k\right) , \tag{3.53}$$

possesses a logarithmic asymptotic expansion with limit $F(x,k)$ *of the form*

$$\sigma_n(x,k) \;=\; F(x,k) \;+\; \sum_{\mu=1} \frac{c_\mu(x,k)}{n^{2\mu}} \;.$$

Furthermore, if we define for all $i \in \mathbb{N}_0$ *and* $0 \le x, k \le 1$ *the sequence of numbers* $y_i = y_i(x,k)$ *by*

$$y_0 \;:=\; x \;, \qquad \text{and for } \; i = 1, 2, \ldots$$

$$y_i \;:=\; 2^i \cdot \sqrt{\left(1 - \sqrt{1 - \left(\frac{y_{i-1}}{2^{i-1}}\right)^2}\right) \Big/ \left(1 + \sqrt{1 - \left(\frac{k y_{i-1}}{2^{i-1}}\right)^2}\right)} ,$$

then for all $i \in \mathbb{N}_0$ *:*

$$y_i \;=\; y_i(x,k) \;=\; \sigma_{2^i}(x,k)$$

with σ_{2^i} *defined in (3.53).*

Remark. For $k = 0$, the iteration rule in Theorem 3.17 reads

$$\begin{aligned} y_i \;&=\; \frac{2^i}{\sqrt{2}} \cdot \sqrt{1 - \sqrt{1 - \left(\frac{y_{i-1}}{2^{i-1}}\right)^2}} \\ &=\; \frac{\sqrt{2} \cdot y_{i-1}}{\sqrt{1 + \sqrt{1 - \left(\frac{y_{i-1}}{2^{i-1}}\right)^2}}} , \end{aligned}$$

a formula which we already know from Lemma 3.12. Thus, as expected,

$$\lim_{i \to \infty} y_i(x,0) \;=\; \arcsin(x) \;.$$

3.4 Numerical Quadrature Rules

Almost every modern exposition of extrapolation methods starts with Romberg's method for "simplified numerical integration", as the original paper from 1955 is entitled (W.ROMBERG [1]). Although this is not quite correct from a historical point of view (cf. the Historical Notes at the end of this book, or C.BREZINSKI [6], D.C.JOYCE [1], G.WALZ [11]), it is at least justified inasmuch as Romberg's paper initiated – in contrast to its predecessors – a huge number of research and publications in this direction, and for example the present booklet would probably not have been written without Romberg's work.

In order to justify the application of Romberg's method or, more general, of convergence acceleration methods for numerical quadrature rules by extrapolation, it is necessary to prove that the quadrature method under consideration (in the Romberg case this is the trapezoidal rule) possesses an asymptotic expansion. This is our aim in the present section; more or less this is always done by a suitable application of Euler's summation formula (Subsection 3.2.2).

3.4.1 The "Classical" Case: Sufficiently Smooth Integrands. In this subsection we shall see that the application of the trapezoidal rule to a sufficiently smooth function f yields a logarithmic asymptotic expansion with limit $\int f$, whose order depends on the smoothness of f.

So, for some interval $[a, b]$ and $r \in \mathbb{N}_0$, let $f \in C^{2r+1}[a, b]$. Suppose we want to approximate the value

$$I_a^b(f) := \int_a^b f(x)\, dx$$

by means of the trapezoidal rule with n subintervals, $n \in \mathbb{N}$, i.e. by the formula

$$T_n(f) := \frac{b-a}{n}\left(\frac{1}{2}f_0 + \sum_{\nu=1}^{n-1} f_\nu + \frac{1}{2}f_n\right), \tag{3.54}$$

where

$$x_\nu := a + \nu \cdot \frac{b-a}{n} \quad \text{and} \quad f_\nu := f(x_\nu) \quad \text{for } \nu = 0, \ldots, n\,.$$

Then the following convergence result holds.

Theorem 3.18: *The sequence of numbers* $\{T_n(f)\}$ *possesses a logarithmic asymptotic expansion with limit* $I_a^b(f)$ *of order* r *; more precisely, we have*

$$T_n(f) = I_a^b(f) + \sum_{\mu=1}^{r} \frac{c_\mu}{n^{2\mu}} + o(n^{-(2r+1)})\,. \tag{3.55}$$

Proof. The assertion follows easily by application of Euler's summation formula (3.25) (with m replaced by n) to the function

$$g(t) = f(a + t\,(b-a)/n)\,,$$

since then

$$g^{(2\mu-1)}(n) - g^{(2\mu-1)}(0) = \frac{(b-a)^{2\mu-1}}{n^{2\mu-1}} \cdot \left(f^{(2\mu-1)}(b) - f^{(2\mu-1)}(a)\right).$$

□

It is remarkable that the coefficients

$$c_\mu = \frac{(b-a)^{2\mu}\,\beta_{2\mu}}{(2\mu)!} \cdot \left(f^{(2\mu-1)}(b) - f^{(2\mu-1)}(a)\right) \tag{3.56}$$

depend only on the value of the derivatives of f at the endpoints of the interval, i.e. the behavior of f in the interior of $[a,b]$ has no influence. This makes it easy to obtain more information on the convergence behavior of the sequence $\{T_n(f)\}$; if, for example, $f'(a) = f'(b)$, then the convergence is (at least) of order n^{-4}. More generally, we obtain as a corollary to Theorem 3.18 the result that for a function $f \in C^{2r+1}(I\!R)$, which is periodic with period $(b-a)$, i.e. if $f(x) = f(x+b-a)$ for all $x \in I\!R$, we have

$$T_n(f) \;=\; I_a^b(f) \;+\; o\big(n^{-(2r+1)}\big)\,. \tag{3.57}$$

This means that the application of an extrapolation method would make no sense in the periodic case.

It is not quite clear at which place in the literature the existence of the asymptotic expansion (3.55) as justification for the applicability of Romberg's method was remarked for the first time. Only one thing is sure: It was *not* in Romberg's original paper itself, where the method is motivated by the fact that the first two steps of it yield well-known quadrature formulae, namely Simpson's rule and Gauß-Cotes' rule (see also E.STIEFEL [1] and Subsection 4.3.1).

So it took at least 5 years until a rigorous analysis of Romberg's method was given. Probably it was F.L.BAUER [1], who did the first step in this direction and showed that the Bernoulli polynomials play an important role. Furthermore we mention the important work of E.STIEFEL [1], J.N.LYNESS & B.J.MCHUGH [1], R.BULIRSCH [1] and R.BULIRSCH & J.STOER [1]. In particular, F.L.BAUER, H.RUTISHAUSER & E.STIEFEL [1] pointed out very clearly that the existence of an asymptotic expansion is necessary for the applicability of a "Romberg type" method; see also H.C.BOLTON & H.I.SCOINS [1,2].

In the booklet by P.C.CHAKRAVARTI [1], many modifications of the asymptotic expansion (3.55) for integrals with weight functions are worked out. However, in most of the cases these weights are smooth functions, such that they affect only the coefficients of the expansion (and not the structure of it).

Besides the trapezoidal rule, the *midpoint rule* is the most prominent numerical quadrature method. Also for this method, the existence of a logarithmic asymptotic expansion can be shown with similar techniques as in the proof of Theorem 3.18; we shall do this even for a generalization of the classical midpoint rule, which was introduced by F.STETTER [1], building on earlier work of D.JAGERMAN [1].

Let ω be a function, which is integrable and nonnegative on the interval $[a,b]$ and has only finitely many zeros there. We shall refer to ω as a weight function, and we are interested in the numerical computation of the integral

$$I_\omega(f) \;:=\; \int_a^b \omega(x)\cdot f(x)\,dx\;.$$

Assuming, without any loss of generality that $I_\omega(1)=1$, we introduce for all $z\in[a,b]$ the function H by

$$y \;=\; H(z) \;:=\; \int_a^z \omega(x)\,dx \quad \text{with } H(b)=1\;.$$

Since H is strictly increasing, the inverse function H^{-1} exists and will be denoted by L:

$$z \;=\; L(y) \;:=\; H^{-1}(y)\;.$$

For $n\in I\!N$ and $\nu=0,\ldots,n-1$, we now fix the *Stetter points*

$$s_\nu \;:=\; n\cdot\int_{\nu/n}^{(\nu+1)/n} L(y)\,dy\;,$$

and define the *generalized midpoint rule* or *Stetter's rule* by

$$M_n(f) \;:=\; \frac{1}{n}\cdot\sum_{\nu=0}^{n-1} f(s_\nu)\;.$$

Note that for $\omega(x) \equiv 1/(b-a)$ we obtain

$$s_\nu \;=\; a + \frac{b-a}{2n} \cdot (2\nu + 1) \,,$$

i.e. $M_n(f)$ reduces to the classical midpoint rule in this case.

It was already indicated in the original paper by F.STETTER [1] that for sufficiently smooth ω and f the error of $M_n(f)$ has convergence order $O(n^{-2})$ for $n \to \infty$. Recently, this result was sharpened by S.GONZALEZ-PINTO, P.GONZALEZ-VERA & J.C.SANTOS [1] inasmuch as the existence of a whole logarithmic asymptotic expansion was shown; the proof of this fact follows, after a nice trick at the beginning, the same ideas as that of our Theorem 3.18, and so we give only a short outline.

Let us define, for $n \in \mathbb{N}$, the functions

$$\alpha(y,n) \;:=\; \int_0^1 L(y + \frac{2z-1}{2n})\, dz \quad \text{and}$$

$$g(y,n) \;:=\; f(\alpha(y,n))$$

on the set

$$\Omega = \Omega(n) := \{(y,n) \mid 0 < y - 1/2n < 1 \;\text{ and }\; 0 < y + 1/2n < 1\} \,.$$

Putting, for $\nu = 0, \ldots, n-1$,

$$y_\nu \;:=\; \frac{2\nu+1}{2n} \,,$$

we recognize that the Stetter points s_ν can also be written as $s_\nu = \alpha(y_\nu, n)$, $\nu = 0, \ldots, n-1$, and therefore

$$\begin{aligned} M_n(f) \;&=\; \frac{1}{n} \cdot \sum_{\nu=0}^{n-1} g(y_\nu, n) \\ &=\; \frac{1}{2n} \cdot \sum_{\nu=0}^{n-1} \left(g(y_\nu, n) + g(y_\nu, -n)\right) \,, \end{aligned}$$

due to the symmetry of g with respect to n. Now, assuming that g (and therefore f) is sufficiently smooth, say $g \in C^{2r+1}(\overline{\Omega})$ for some $r \in I\!N$, Taylor's formula yields a representation of the form

$$g(y_\nu, n) + g(y_\nu, -n) = \sum_{j=0}^{r} \frac{\gamma_j}{n^{2j}} + o(n^{-(2r+1)}) \text{ for } n \to \infty$$

with certain coefficients γ_j, independent of n. From this point on, straightforward considerations, using again Euler's summation formula, lead us to the following result.

Theorem 3.19 (S.GONZALEZ-PINTO, P.GONZALEZ-VERA & J.C.SANTOS [1]): *Suppose that there is a natural number $N > 0$, such that the function $g(y,n)$ from above possesses bounded partial derivatives up to the order $2r+1$ on $\overline{\Omega(n)}$ for all $n \geq N$.*
Then the generalized midpoint rule possesses a logarithmic asymptotic expansion with limit $I_\omega(f)$ of order r; we have

$$M_n(f) = I_\omega(f) + \sum_{\mu=1}^{r} \frac{c_\mu}{n^{2\mu}} + o(n^{-(2r+1)})$$

with certain coefficients c_μ, independent of n.

Remark. If the integral $n \int_{\nu/n}^{(\nu+1)/n} L(y)dy$, which had us defined the Stetter points s_ν, is approximated by the (classical) midpoint rule, i.e. by $L((2\nu+1)/2n)$, we recover *Jagerman's rule*

$$\widetilde{M}_n(f) := \frac{1}{n} \cdot \sum_{\nu=0}^{n-1} f(j_\nu)$$

with

$$j_\nu := L((2\nu+1)/2n)$$

for $\nu = 0, \ldots, n-1$. In contrast to Stetter's rule, which is known to be exact for polynomials $p \in \Pi_1$ in general, the error of Jagerman's rule vanishes for arbitrary weight functions only if $p \in \Pi_0$. However, also for Jagerman's rule it is possible to prove the existence of a logarithmic asymptotic expansion and therefore to apply suited extrapolation procedures, cf. S.GONZALEZ-PINTO, P.GONZALEZ-VERA & J.C.SANTOS [1] and T.HÅVIE [2].

Quite another understanding of the term asymptotic expansion forms the basis of the work by A. Sidi on the numerical computation of infinite integrals. In a series of papers, a nice compendium of which is given in A.SIDI [4], he develops extrapolation algorithms for the computation of integrals of the form

$$\int_a^\infty f(x)\,dx \qquad \text{with } a \geq 0\,,$$

mainly under the assumption that the remainder term

$$\int_a^\infty f(x)\,dx - \int_a^\xi f(x)\,dx \;=\; \int_\xi^\infty f(x)\,dx$$

possesses a well-structured asymptotic expansion as $\xi \to \infty$. For example, in D.LEVIN & A.SIDI [1] it is worked out that for a certain class of integrands f asymptotic expansions of the form

$$\int_\xi^\infty f(x)\,dx \;=\; \sum_{\mu=0}^{m-1} \xi^{\rho_\mu} f^{(\mu)}(\xi) \sum_{i=0}^{\infty} \gamma_{\mu i}\xi^{-i} \quad \text{for } \xi \to \infty$$

exist, where $\rho_\mu \leq \mu + 1$ are integers depending on f . On the basis of this knowledge, different types of extrapolation methods can be applied for the numerical computation of the infinite integral.

Let us return now to the trapezoidal rule and investigate in the next subsections what happens if the integrand has a singularity in the integration range.

3.4.2 Integrands with an Endpoint Singularity. If there are singularities of f or its derivatives within the range of integration, the nice asymptotic expansion given in Theorem 3.18 is no longer valid. However, soon after Romberg's original work had

Corollary 3.22: *Let the functions* f_α *and* g_β *be as in (3.63), and consider a function* w *, which possesses everywhere in* $[0,b]$ *a series representation of the form*

$$w(x) = \sum_{j=0}^{\infty} a_j x^j .$$

Then the following asymptotic expansions for the trapezoidal rule hold:

$$T_n(f_\alpha \cdot w) = I_0^b(f_\alpha \cdot w) + \sum_{\mu=1} \frac{\gamma_\mu}{n^{\alpha+\mu}} + \sum_{\mu=1} \frac{c_\mu}{n^{2\mu}}$$

and

$$T_n(g_\beta \cdot w) = I_0^b(g_\beta \cdot w) + \sum_{\mu=1} \frac{\gamma_\mu}{n^{\beta+\mu}} + \\ + \sum_{\mu=1} \frac{\tilde{\gamma}_\mu}{n^{\beta+\mu}\,(\log(1/n))^{-1}} + \sum_{\mu=1} \frac{c_\mu}{n^{2\mu}} .$$

Proof. An application of the same arguments as in the proof of Theorem 3.21 to each term of the resulting series

$$f_\alpha(x)\cdot w(x) = \sum_{j=0}^{\infty} a_j x^j f_\alpha(x) \quad \text{and} \quad g_\beta(x)\cdot w(x) = \sum_{j=0}^{\infty} a_j x^j g_\beta(x) ,$$

i.e. to integrands of the form

$$x^{j+\alpha} \quad \text{and} \quad x^{j+\beta}\cdot\log(x) , \tag{3.64}$$

$j \in I\!N_0$, immediately gives us the desired result. □

Further possible extensions of Theorem 3.21 are also obvious. But we want now to leave this field and turn our attention to the case where the integrand itself has a singularity in (one of) the endpoints of the integration range. As before, there exist at least four completely different proofs of the corresponding main result (Theorem 3.23) on the existence of an asymptotic expansion, but each of them fills at least one whole paper and is therefore too long to be presented here. For example, the idea behind the proof due to I.Navot [1] is very similar to that of L. Fox for Theorem 3.21, which we presented above. But the precise elaboration of this idea, i.e. splitting of the interval $[a, b]$ into $[a, x_1]$ and $[x_1, b]$ and handling their contributions to the error terms separately is more complicated here.

Let us first assume that our integrand $h_{\alpha,0}$ is now of the form

$$h_{\alpha,0}(x) \;=\; x^{\alpha} \cdot w(x)$$

(the 0 being inserted for later generalization) with w sufficiently smooth and $-1 < \alpha \leq 0$, and that we want to approximate the integral

$$I_0^b(h_{\alpha,0}) \;=\; \int_0^b h_{\alpha,0}(x)\, dx$$

by means of the the trapezoidal rule, excluding the evaluation of the lower endpoint. Then the following theorem holds.

Theorem 3.23 (J.N.Lyness & B.W.Ninham [1], I.Navot [1], P.C.Waterman, J.M.Yos & R.J.Abodeely [1], H.Rutishauser [1], P.C.Chakravarti [1]): *With the assumptions and notations from above, the sequence of modified trapezoidal values*

$$T_n'(h_{\alpha,0}) \;:=\; \frac{b}{n}\left(\sum_{\nu=1}^{n-1} h_{\alpha,0}(x_\nu) + \frac{1}{2}\, h_{\alpha,0}(b)\right) \tag{3.65}$$

($x_\nu := \frac{\nu \cdot b}{n}$) possesses the logarithmic asymptotic expansion

$$T_n'(h_{\alpha,0}) \;=\; I_0^b(h_{\alpha,0}) \;+\; \sum_{\mu=1} \frac{\gamma_\mu}{n^{\alpha+\mu}} \;+\; \sum_{\mu=1} \frac{c_\mu}{n^{2\mu}} \tag{3.66}$$

with the coefficients

$$\gamma_\mu \;=\; \frac{\zeta(-(\mu-1)-\alpha)\cdot b^\mu}{(\mu-1)!}\cdot w^{(\mu-1)}(0) \tag{3.67}$$

and

$$c_\mu \;=\; \frac{\beta_{2\mu}\cdot b^{2\mu}}{(2\mu)!}\cdot h_{\alpha,0}^{(2\mu-1)}(b) \tag{3.68}$$

where ζ *is Riemann's zeta-function and* $\beta_{2\mu}$ *denotes the Bernoulli numbers, see Section 3.2.*

We note that the expansion (3.66) coincides with that in Corollary 3.22, which was shown to be valid for $0 < \alpha < 1$. But what about the case $\alpha = 0$, meaning that $h_{\alpha,0} = w$ and the integrand has no singularity at all?

In this case we use the well-known identities

$$\zeta(0) = -\frac{1}{2}\,, \quad \zeta(1-2j) = -\frac{\beta_{2j}}{2j}\,, \quad \zeta(-2j) = 0$$

for $j \in \mathbb{N}$, which give us in combination with (3.67) the values of the coefficients

$$\gamma_1 \;=\; -\frac{b}{2}\cdot w(0)$$

and, for $\mu \geq 2$,

$$\gamma_\mu \;=\; \begin{cases} -\dfrac{\beta_\mu\, b^\mu}{\mu!}\cdot w^{(\mu-1)}(0)\,, & \text{if } \mu \text{ is even,} \\ 0\,, & \text{if } \mu \text{ is odd.} \end{cases}$$

Inserting this result into (3.66) yields

$$T_n'(w) \;=\; I_0^b(w) \;-\; \frac{b}{2n}w(0) \;+\; \sum_{\mu=1} \frac{\tilde{c}_\mu}{n^{2\mu}}$$

with the coefficients

$$\tilde{c}_\mu = \gamma_{2\mu} + c_\mu = \frac{\beta_{2\mu} \cdot b^{2\mu}}{(2\mu)!} \cdot \left(w^{(2\mu-1)}(b) - w^{(2\mu-1)}(0)\right) ,$$

that is, we recover the usual asymptotic expansion (3.55) for smooth integrands as special case of the expansion (3.66), and all together we have shown the validity of (3.66) for

$$-1 < \alpha < 1 .$$

Already in 1961, H. Rutishauser was aware of the structure of the expansion (3.66) at least for the case

$$w \equiv 1 , \; b = 1 \; \text{ and } \; \alpha \in \mathbb{C}, \; -1 < \mathrm{Re}\alpha < 0 , \tag{3.69}$$

and used this for numerical tests of the Riemann hypothesis (E.STIEFEL [1], see also H.RUTISHAUSER [1]). The idea is as follows: With the limitations given in (3.69), the first coefficient of the expansion (3.66) (and this is the only one which belongs to a power of n with real part smaller than 1) reads

$$\gamma_1 = \zeta(-\alpha) .$$

Therefore, the sequence $\{T_n'(x^{-\alpha})\}$ converges faster than n^{-1} , if and only if $-\alpha$ is a root of the zeta-function. Rutishauser's numerical tests illustrated very nicely that this kind of convergence happens only if $\mathrm{Re}\alpha = -\frac{1}{2}$, and that would mean that Riemann's hypothesis is true.

With the same methods as in the proof of Theorem 3.23, or just by application of the Theorem itself, one can prove the following Corollary 3.24, which extends the result to functions having a singularity in both endpoints of the interval.

Without any loss of generality, we restrict ourselves now to $[0,1]$, i.e. we approximate the integral

$$I_0^1(h_{\alpha,\beta}) = \int_0^1 h_{\alpha,\beta}(x)\, dx \tag{3.70}$$

of the function

$$h_{\alpha,\beta}(x) = x^\alpha \cdot (1-x)^\beta \cdot w(x)$$

with $-1 < \alpha , \beta < 0$ and w as above.

Corollary 3.24 (J.N.LYNESS & B.W.NINHAM [1], P.C.CHAKRAVARTI [1]): *With the assumptions and notations from above, the sequence of modified trapezoidal values*

$$T_n''(h_{\alpha,\beta}) := \frac{1}{n} \cdot \sum_{\nu=1}^{n-1} h_{\alpha,\beta}(x_\nu) \tag{3.71}$$

($x_\nu := \frac{\nu}{n}$) possesses the logarithmic asymptotic expansion

$$T_n''(h_{\alpha,\beta}) = I_0^1(h_{\alpha,\beta}) + \sum_{\mu=1} \frac{\gamma_\mu}{n^{\alpha+\mu}} + \sum_{\mu=1} \frac{\delta_\mu}{n^{\beta+\mu}} \tag{3.72}$$

with the coefficients

$$\gamma_\mu = \frac{\zeta(-(\mu-1)-\alpha)}{(\mu-1)!} \cdot h_{0,\beta}^{(\mu-1)}(0) \tag{3.73}$$

and

$$\delta_\mu = \frac{(-1)^{\mu-1} \cdot \zeta(-(\mu-1)-\beta)}{(\mu-1)!} \cdot h_{\alpha,0}^{(\mu-1)}(1)\,. \tag{3.74}$$

Proof. Left to the reader (see Problem P 3.10). □

If $\alpha < -1$ or $\beta < -1$, the integral in (3.70) is no longer a convergent one, and the methods from above cannot be applied. However, using generalized function theory in the form presented by M.J.LIGHTHILL [1] it was shown by B.W.NINHAM [1] that also for $\alpha < -1$ and/or $\beta < -1$ it makes sense to write down an asymptotic expansion like (3.72).

Moreover, this provides a very efficient method for the numerical evaluation of the corresponding integral (or, to be more precise, for its Hadamard finite part, which is the analytic continuation of the Riemann integral (3.70), see B.W.NINHAM [1]). As a particular result, Ninham illustrated numerically that the equality

$$\int_0^1 x^\alpha \cdot (1-x)^\beta \, dx = \frac{\alpha! \cdot \beta!}{(\alpha+\beta+1)!}$$

remains valid *for all* $\alpha, \beta \in I\!R$, $\alpha + \beta \neq -1$.

So far we have only considered integrands having an algebraic or branch endpoint singularity, but it is easy now to obtain an asymptotic expansion for integrals over functions with a logarithmic endpoint singularity (where the asymptotic expansion is – maybe confusingly – no longer a logarithmic one, cf. the remark after Theorem 3.21!) by differentiation of both sides in (3.72) with respect to one of the exponents, say, to α.

Corollary 3.25 (J.N.Lyness & B.W.Ninham [1], I.Navot [2]): *Consider evaluation of the integral*

$$I_0^1(\tilde{h}_{\alpha,\beta}) := \int_0^1 x^\alpha \cdot (1-x)^\beta \cdot \log(x) \cdot w(x)\,dx$$

by the modified trapezoidal rule T_n'' *as defined in (3.71).*
Then the sequence $\{T_n''(\tilde{h}_{\alpha,\beta})\}$ *possesses an asymptotic expansion of the form*

$$\begin{aligned} T_n''(\tilde{h}_{\alpha,\beta}) &= I_0^1(\tilde{h}_{\alpha,\beta}) + \sum_{\mu=1} \frac{\tilde{\gamma}_\mu}{n^{\alpha+\mu}} + \\ &\quad + \sum_{\mu=1} \frac{\gamma_\mu}{n^{\alpha+\mu}\,(\log(1/n))^{-1}} + \sum_{\mu=1} \frac{\tilde{\delta}_\mu}{n^{\beta+\mu}} . \end{aligned} \tag{3.75}$$

Proof. Differentiation of the asymptotic expansion (3.72) with respect to α immediately yields the desired result, since

$$\frac{d}{d\alpha} h_{\alpha,\beta}(x) = \tilde{h}_{\alpha,\beta}(x)$$

and

$$\frac{d}{d\alpha} \frac{\gamma_\mu}{n^{\alpha+\mu}} = \frac{\tilde{\gamma}_\mu}{n^{\alpha+\mu}} + \frac{\gamma_\mu}{n^{\alpha+\mu}\,(\log(1/n))^{-1}}$$

for all μ. □

Most of the just mentioned results are also summarized in J.N.Lyness & E.deDoncker [1], partly supplied with proofs different from those above, and in PH.RABINOWITZ [1]. Furthermore, there exist other approaches which use a transformation of the integrand into another form having no singularities any more; a combination of this idea with Romberg's method is discussed in D.B.HUNTER [1].

It is most probably clear now to the reader that there do exist many more modifications of the classical summation formula to cope with many different kinds of endpoint sigularities, which can of course not all be included in this booklet; for example, results are known corcerning quasi-step discontinuities (cf. I.NAVOT [3]) or essential endpoint singularities (P.C.WATERMAN, J.M.YOS & R.J.ABODEELY [1]) of the integrand. Also, the situation where the integrand itself is smooth within the interval, but has a singularity in the complex plane close to the range of integration, which might eventually cause numerical instabilities, was studied independently by I.NAVOT [4] and by B.W.NINHAM & J.N.LYNESS [1]. Furthermore, the problem of evaluating a complex contour integral was considered by J.N.LYNESS & L.M.DELVES [1].

3.4.3 Integrands with a Singularity in the Interior of the Integration Range. In this section we treat the case of integrands having a singularity, say, a simple pole, *within* the range of integration; that is, we want to approximate the Cauchy principal value integral

$$CI_0^1(f) \;:=\; \lim_{\varepsilon\to 0+} \left\{ \int_0^{c-\varepsilon} f(x)\,dx \;+\; \int_{c+\varepsilon}^1 f(x)\,dx \right\}\,, \qquad (3.76)$$

where the integrand is supposed to have a first order pole for $x = c$, $0 < c < 1$. We assume that f is of the form

$$f(x) \;=\; (x-c)^{-1} w(x) \qquad (3.77)$$

with w being sufficiently smooth everywhere in $[0,1]$ (this will be stated more precisely in Theorem 3.27) and $w(c) \neq 0$. In this case, the number

$$\varrho := \varrho(f) := \lim_{x\to c} (x-c)\cdot f(x) \qquad (3.78)$$

is finite and different from zero. Moreover, we set

$$\tilde{\varrho} := \tilde{\varrho}(f) := \lim_{x \to c} \frac{d}{dx} \left[(x-c) \cdot f(x) \right] .$$

It was shown independently by D.B.HUNTER [2] and by J.N.LYNESS [5] that once again the sequence of suitably modified trapezoidal values possesses a logarithmic asymptotic expansion of the usual form; "suitable modification" means in this context that one has to subtract a certain correction term from the original trapezoidal values.

The following derivation of Theorem 3.27, the main result of this subsection, is more or less a mixture of the two approaches given by Hunter and Lyness; we first define the *subtraction function*

$$\Psi(x) \; := \; \begin{cases} 0, & \text{if } x - c \in \mathbb{Z}, \\ \pi \varrho \cot(\pi(x-c)), & \text{if } x - c \notin \mathbb{Z}; \end{cases} \tag{3.79}$$

note that the second term in this definition, say Ψ_1, can also be written as

$$\pi i \varrho \left(1 - \frac{2}{1 - \exp(2\pi i (x-c))} \right) . \tag{3.80}$$

We list some properties of the function Ψ, which will be useful in the proof of Theorem 3.27.

Lemma 3.26: *a) The function Ψ_1 has a simple pole at $x = c$, the residue at this point being*

$$\operatorname{Res}(\Psi_1, c) \; = \; \varrho .$$

b) For all $x \in \mathbb{R}$, $\Psi(x+1) = \Psi(x)$.

c) Ψ is an odd function with respect to c and to $c + 1/2$, i.e. for all $x \in \mathbb{R}$

$$\begin{aligned} \Psi(c+x) \; &= \; -\Psi(c-x) \quad \text{and} \\ \Psi(c+1/2+x) \; &= \; -\Psi(c+1/2-x) . \end{aligned}$$

d) For each real number ξ *, it is*

$$CI_{\xi}^{\xi+1}(\Psi) = 0 .$$

e) The trapezoidal rule T_n *, applied to* Ψ *, yields*

$$T_n(\Psi) = \begin{cases} 0, & \text{if } nc \in \mathbb{N}, \\ -\pi\varrho\cot(\pi nc), & \text{if } nc \notin \mathbb{N}. \end{cases} \tag{3.81}$$

Proof. Properties a), b) and c) can be read off directly from the definition of Ψ .
To prove d), we only have to use properties b) and c) to find

$$CI_{\xi}^{\xi+1}(\Psi) = CI_{c-\frac{1}{2}}^{c+\frac{1}{2}}(\Psi) = 0$$

for each $\xi \in \mathbb{R}$.

We now show the validity of (3.81). If $nc \in \mathbb{N}$, then the singular point c belongs to the abscissae of T_n , and the result $T_n(\Psi) = 0$ follows directly from property c) and the fact that $\Psi(c) = 0$.

Now suppose that nc is not an integer. We use the well-known formula (see Problem P.3.11)

$$T_n(e_k) = \begin{cases} 0, & \text{if } k/n \notin \mathbb{Z}, \\ 1 & \text{if } k/n \in \mathbb{Z}, \end{cases} \tag{3.82}$$

where k is an arbitrary integer and e_k denotes the auxiliary function $e_k(x) := \exp(2\pi ikx)$.
With some $\varepsilon > 0$, ε small enough, we apply the trapezoidal rule T_n to the function

$$\begin{aligned} \Psi(x+i\varepsilon) &= \pi i\varrho\left(1 - \frac{2}{1-\exp(-2\pi\varepsilon)\exp(2\pi i(x-c))}\right) \\ &= \pi i\varrho\left(1 - 2\sum_{k=0}^{\infty}\exp(-2\pi k\varepsilon)\exp(2\pi ik(x-c))\right) \end{aligned}$$

and we obtain, using the Relation (3.82),

$$T_n(\Psi(\cdot\ + i\varepsilon)) = \pi i \varrho \left(1 - 2\sum_{\nu=0}^{\infty} \exp(-2\pi\nu n\varepsilon)\exp(-2\pi i\nu nc)\right)$$

$$= \pi i \varrho \left(1 - \frac{2}{1-\exp(-2\pi n\varepsilon)\exp(-2\pi inc)}\right).$$

Since nc is not an integer, we may let ε in this last relation tend to zero, which gives us

$$T_n(\Psi) = \pi i \varrho \left(1 - \frac{2}{1-\exp(-2\pi inc)}\right) = \pi\varrho\cot(-\pi nc),$$

and since cot is an odd function, we have proved the last statement of Lemma 3.26. □

Now, consider the integrand function f to be given as in (3.77) for all $x \neq c$, and set $f(c) := w'(c) = \tilde{\varrho}$. Then f is well-defined everywhere in $[0,1]$, and so is the application of the trapezoidal rule T_n to f. However, these (unmodified) trapezoidal values bear no relation to the values of the Cauchy principal value integrals $CI(f)$ (see Table 3.2). But the following Theorem 3.27 tells us how one can slightly modify the values T_n in order to obtain a new sequence $\{CT_n\}$, which possesses again a logarithmic asymptotic expansion with the desired limit $CI_0^1(f)$.

Theorem 3.27 (D.B.Hunter [2], J.N.Lyness [5]): *Let the function f for all $x \in [0,1]$ be of the form*

$$f(x) = \begin{cases} (x-c)^{-1}w(x), & x \neq c, \\ w'(c), & x = c, \end{cases} \tag{3.83}$$

with $w \in C^{2r+1}[0,1]$ for some $r \in \mathbb{N}_0$ and $w(c) \neq 0$, and adopt the previous notations. Furthermore, define for all $n \in \mathbb{N}$ the modified trapezoidal values $CT_n(f)$ through

$$CT_n(f) := T_n(f) + \begin{cases} 0, & \text{if } nc \in \mathbb{N}, \\ \pi\varrho\cot(\pi nc), & \text{if } nc \notin \mathbb{N}. \end{cases} \tag{3.84}$$

Then the sequence $\{CT_n(f)\}$ *possesses the following logarithmic asymptotic expansion of order* r *:*

$$CT_n(f) \;=\; CI_0^1(f) \,+\, \sum_{\mu=1}^{r} \frac{c_\mu}{n^{2\mu}} \,+\, o(n^{-(2r+1)})\,, \tag{3.85}$$

where

$$c_\mu \;=\; \frac{(b-a)^{2\mu}\,\beta_{2\mu}}{(2\mu)!}\cdot\left(f^{(2\mu-1)}(1) - f^{(2\mu-1)}(0)\right) \tag{3.86}$$

for all $\mu \in \mathbb{N}_0$.

Proof. We apply the integral operator CI_0^1 to the difference function $h := f - \Psi$. Since $h \in C^{2r+1}[0,1]$, it follows that

$$CI_0^1(f) \;=\; CI_0^1(f) - CI_0^1(\Psi) \;=\; CI_0^1(h) \;=\; I_0^1(h)\,, \tag{3.87}$$

i.e. the Cauchy principal value integral over the function f coincides with the standard integral over the function h .

Moreover, since Ψ is a periodic function with period length 1 , we find that $\Psi^{(\mu)}(0) = \Psi^{(\mu)}(1)$ for all $\mu \in \mathbb{N}_0$, and so

$$h^{(\mu)}(1) - h^{(\mu)}(0) \;=\; f^{(\mu)}(1) - f^{(\mu)}(0) \tag{3.88}$$

for all $\mu \in \mathbb{N}_0$.
Now we apply the result of Theorem 3.18 to the difference function h and obtain the relation

$$T_n(h) \;=\; I_0^1(h) \,+\, \sum_{\mu=1}^{r} \frac{c_\mu}{n^{2\mu}} \,+\, o(n^{-(2r+1)}) \tag{3.89}$$

with the coefficients

$$c_\mu \;=\; \frac{(b-a)^{2\mu}\,\beta_{2\mu}}{(2\mu)!}\cdot\left(h^{(2\mu-1)}(1) - h^{(2\mu-1)}(0)\right)\,.$$

Using (3.87) and (3.88), it follows that the right hand side of (3.85) is identical with that of (3.89).

It remains to prove that also the left hand members of these two relations coincide, i.e. that

$$T_n(h) \;=\; CT_n(f)\,. \tag{3.90}$$

If $nc \in I\!N$, then one of the function values required for the evaluation of T_n concides with the singular point c. Therefore we obtain, using the notation

$$\tilde{T}_n(f) \;:=\; \frac{1}{n}\Big(\frac{1}{2}\,f(0) + \sum_{\substack{\nu=1\\ \nu\neq nc}}^{n-1} f\Big(\frac{\nu}{n}\Big) + \frac{1}{2}\,f(1)\Big)$$

(and analogously for the other functions):

$$\begin{aligned} T_n(h) \;&=\; \tilde{T}_n(h) + \frac{1}{n}\cdot h(c) \;=\; \tilde{T}_n(f) - \tilde{T}_n(\Psi) + \frac{1}{n}\cdot h(c) \\ &=\; \tilde{T}_n(f) - \tilde{T}_n(\Psi) + \frac{1}{n}\cdot f(c) \\ &=\; T_n(f) - \tilde{T}_n(\Psi) \;=\; CT_n(f) - \tilde{T}_n(\Psi)\,, \end{aligned}$$

where we have used that $f(c) = h(c)(= w'(c))$. Since $\Psi(c) = 0$, we may write $T_n(\Psi)$ instead of $\tilde{T}_n(\Psi)$, and because $T_n(\Psi) = 0$, we have proved (3.90) in the case $nc \in I\!N$.

Finally, if $nc \notin I\!N$, then the singular point c does not belong to the required abscissae of h resp. f, and it follows that

$$T_n(h) \;=\; T_n(f) - T_n(\Psi) \;=\; CT_n(f)\,,$$

according to (3.81). This completes the proof of Theorem 3.27. □

Example. We illustrate the strength of this result with a numerical example. Consider the function

$$f_c(x) := \frac{\exp(-x)}{x-c}, \tag{3.91}$$

which is of the type under consideration: it has a simple pole for $x = c$, the residue in this point being $\varrho(f_c) = \exp(-c)$. Moreover, one can show (cf. D.B.HUNTER [2]) that the Cauchy principal value integral over f_c can be expressed in terms of the exponential integral; we have

$$CI_0^1(f_c) = -\exp(-c) \cdot (Ei(c) + E_1(1-c)),$$

in the notation of M.ABRAMOWITZ & I.STEGUN [1]. For $c = 0.2$, we therefore find the true value $CI_0^1(f_{0.2}) = 0.41851$.

n	T_n	CT_n	E_n	$E_{n/2}/E_n$
1	−2.27008	1.27014	0.851 $e(\ 0)$	
2	−0.12415	0.71158	0.293 $e(\ 0)$	2.90589
4	4.04664	0.50642	0.879 $e(-1)$	3.33361
8	1.27804	0.44230	0.238 $e(-1)$	3.69388
16	−3.11560	0.42462	0.611 $e(-2)$	3.89520
32	−0.41568	0.42004	0.153 $e(-2)$	3.97289
64	3.95911	0.41887	0.381 $e(-2)$	4.04016
128	1.25434	0.41859	0.921 $e(-3)$	4.13499

Table 3.2. Approximation of $CI_0^1(f_{0.2})$ (see Text)

Table 3.2 shows the output of the classical trapezoidal rule (denoted as T_n), the values of the modified trapezoidal rule (3.84) (CT_n), and the approximation errors of the latter ones in absolute value (E_n). Quite obviously, the numbers T_n bear no relationship with the true value of $CI_0^1(f_{0.2})$ given above, while the CT_n's approximate it quite well (but note that $nc \neq I\!N$ for all n, hence the trapezoidal rule T_n does not even "notice" that the function under consideration possesses a pole!).

Moreover, from the entries of the last column, the quotients $\frac{E_{n/2}}{E_n}$, it can be recognized that the sequence of these quotients converges to 4 , in accordance to the result of Theorem 3.27.

Remark. Since the modification given by (3.84) is a completely linear process, one can with the same method also handle functions which have more than one pole within the integration range. That is, if the function f possesses the simple poles $c_1, \ldots, c_k$ in $[0,1]$, then the subtraction function Ψ from (3.79) has to be replaced by the sum $\sum_{j=1}^{k} \Psi_j$, each Ψ_j being of the form

$$\Psi_j(x) := \begin{cases} 0, & \text{if } x - c_j \in \mathbb{Z}, \\ \pi \varrho_j \cot(\pi(x - c_j)), & \text{if } x - c_j \notin \mathbb{Z} , \end{cases}$$

with ϱ_j defined as in (3.78), $j = 1, \ldots, k$. The modification process in (3.84) then reads

$$CT_n(f) := T_n(f) + \sum_{j=1}^{k} R_{n,j}(f)$$

where, for $j = 1, \ldots, k$,

$$R_{n,j}(f) = \begin{cases} 0, & \text{if } nc_j \in \mathbb{N}, \\ \pi \varrho_j \cot(\pi n c_j), & \text{if } nc_j \notin \mathbb{N}. \end{cases}$$

Furthermore, by combining Theorem 3.27 with the various results of Subsection 3.4.2, it is possible to treat also Cauchy principal value integrals of functions having an endpoint singularity. As an example for that, we present the following result from J.N.LYNESS [5], which follows now from a combination of Corollary 3.24 with Theorem 3.27.

Corollary 3.28: *Let $0 < c < 1$, and assume that the function f is for all $x \in [0,1]$ given by*

$$f(x) \;=\; \begin{cases} x^\alpha(1-x)^\beta(x-c)^{-1}w(x), & x \neq c, \\ \lim\limits_{t\to c}\left(\frac{d}{dt}\left[t^\alpha(1-t)^\beta w(t)\right]\right), & x = c, \end{cases} \tag{3.92}$$

where $-1 < \alpha, \beta < 0$, $w \in C^\infty[0,1]$, and $w(c) \neq 0$.

Furthermore, with $\varrho := c^\alpha(1-c)^\beta w(c)$, define for all $n \in I\!N$ the modified trapezoidal values $CT_n(f)$ through

$$CT_n(f) \;:=\; T_n''(f) \;+\; \begin{cases} 0, & \text{if } nc \in I\!N, \\ \pi\varrho\cot(\pi n c), & \text{if } nc \notin I\!N. \end{cases} \tag{3.93}$$

Then the sequence $\{CT_n(f)\}$ possesses a logarithmic asymptotic expansion of the following type:

$$CT_n(f) \;=\; CI_0^1(f) \;+\; \sum_{\mu=1} \frac{\gamma_\mu}{n^{\alpha+\mu}} \;+\; \sum_{\mu=1} \frac{\delta_\mu}{n^{\beta+\mu}}\,.$$

If the location of the pole c is not known, the above-mentioned methods cannot be applied. In D.B.Hunter & H.V.Smith [1] it is indicated how to modify the formulas in order to detect an previously unknown pole.

3.4.4 Multivariate Integration. We now make some remarks on what is sometimes called "cubature and hypercubature", that is, the numerical evaluation of integrals over bivariate or multivariate functions. Of course we shall only treat that part of the theory which is concerned with asymptotic expansions. Although almost all results look very much like the ones for univariate functions presented in the previous subsections, it must be said that the proofs are sometimes quite different and involve sophisticated analytical methods.

Here is a brief overview over the existing results in this field: The story probably begins with a paper by J.N.Lyness & B.J.McHugh [1] from 1963, which treats the case of smooth integrands over the $s-$ dimensional cube. This theory was worked out further in a little series of papers by J.N.LYNESS [1], J.N.LYNESS & B.J.MCHUGH [2], and independently also by C.T.BAKER & G.S.HODGSON [1].

Later on, the case of integrands having a singularity at a vertex (J.N.LYNESS [2]) or at one (J.N.LYNESS & E.DEDONCKER [1], A.SIDI [1]) or two (J.N.LYNESS & E.DEDONCKER [2]) boundary lines of the cube resp. the square were treated, and G.MONEGATO & J.N.LYNESS [1] considered the problem of evaluating the Cauchy principal value integral $\int f(x,y)/|x-y|\,dx\,dy$. The results in all of these cases are, with a self-evident change of notation, very similar to those of the univariate case (but keep in mind the above-stated remark on the proofs!). As a typical example, we present the followig theorem concerning the integral

$$I^{(2)}(\Gamma_\alpha) \;:=\; \int\limits_S \Gamma_\alpha(x,y)dx\,dy \ ,$$

over the unit square $S := \{(x,y) \in \mathbb{R}^2; 0 \le x,y \le 1\}$, where the function Γ_α as defined below is singular at one boundary line of S .

Theorem 3.29 (A.SIDI [1], J.N.LYNESS & E.DEDONCKER [2]): *Consider a function $f(x,y) \in C^{2r}(S)$, $r \in \mathbb{N}$, and choose some $\alpha \in \mathbb{R}$, $\alpha > -1$. Now, apply the bivariate trapezoidal rule $T_n^{(2)}$ to the function $\Gamma_\alpha(x,y) := x^\alpha \cdot f(x,y)$, i.e. evaluate*

$$T_n^{(2)}(\Gamma_\alpha) \;:=\; \frac{1}{n}\left(\sum_{\nu=1}^{n-1}\left(\frac{\nu}{n}\right)^\alpha \cdot T_n\big(f\big(\tfrac{\nu}{n},\cdot\big)\big)\right) + \frac{1}{2n}T_n\big(f(1,\cdot)\big)\,, \tag{3.94}$$

where T_n denotes the usual univariate trapezoidal rule. Then the logarithmic asymptotic expansion

$$T_n^{(2)}(\Gamma_\alpha) \;=\; I^{(2)}(\Gamma_\alpha) + \sum_{\mu=1}^{r-1}\frac{c_\mu}{n^{2\mu}} + \sum_{\mu=1}^{2r}\frac{\gamma_\mu}{n^{\alpha+\mu}} + O(n^{-2r}) \tag{3.95}$$

for $n \to \infty$ is valid.

If we take as integration range the unit simplex

$$\Big\{(x_1,\ldots,x_s) \in \mathbb{R}^s \;\Big|\; x_\nu \ge 0, \nu = 1,\ldots,s\,;\; x_1 + \cdots x_s \le 1\Big\} \tag{3.96}$$

instead of the unit cube, the results do not change essentially in shape, but once again new methods for the proofs are involved. The theory of asymptotic expansions for numerical integration over the simplex was founded in J.N.LYNESS & K.K.PURI [1], J.N.LYNESS [3,4], E.DEDONCKER [1] (smooth integrands), and further developed by J.N.Lyness & G.Monegato [1] and A.Sidi [1] (integrands with singularities).

Finally we mention a paper by K.Georg [1], in which a new method for the numerical calculation of two-dimensional integrals over a surface in $\mathbb{R}^3$ is introduced. The existence of an asymptotic expansion was only conjectured there, but could be proved later on by P.Verlinden & R.Cools [1]. Moreover, quite recently the existence of logarithmic asymptotic expansions for quadrature rules over curved surfaces was investigated in J.N.Lyness [6].

3.5 Divided Differences and Numerical Differentiation. The E-Method

In contrast to numerical integration, the problem of numerical differentiation of functions is rather poorly treated in the literature. A reason for that might be that in this context always very direct methods are used, which have (or seem to have) not much theory behind themselves. However, we will show in the following that these methods, which are all based on more or less obvious modifications of the classical divided differences, also produce approximating sequences with an logarithmic asymptotic expansion. Therefore, extrapolation methods can be applied.

Moreover, we will study generalized divided differences, which were introduced by T.Popovociu [1] and further investigated by S.Karlin [1], G.Mühlbach [1,3,4] and G.Walz [12,13]; we will prove the existence of an asymptotic expansion also for these generalized divided differences, if the underlying function space is a Polya space.

Furthermore we shall present the so-called E-method, which is a method for the condensation of certain asymptotic expansions, and show its usefulness by means of a specific example, namely the numerical differentiation of a holomorphic function via Cauchy's Theorem.

3.5.1 Classical Divided Differences. In Subsection 3.1.3 we had already encountered an example for the approach we are going to present now: It was shown that the application of two specific first-order divided differences, the forward difference and the central difference, to a function g produces asymptotic expansions with limit function g'. In the following we shall generalize this result of H.RUTISHAUSER [1] considerably. We first have to learn some elementary facts on divided differences:

Definition 3.30: For some $\nu \in \mathbb{Z}$ and $m \in \mathbb{N}_0$, let there be given a set of pairwise distinct numbers $\{x_\nu, \ldots, x_{\nu+m}\}$ and a function g, which is defined at these points.

Then the $m-$ th order divided difference $\Delta(x_\nu, \ldots, x_{\nu+m}; g)$ of g with respect to the points $x_\nu, \ldots, x_{\nu+m}$ is defined recursively by

$$\Delta(x_j; g) := g(x_j) \qquad \text{for } j = \nu, \ldots, \nu + m,$$

$$\Delta(x_j, \ldots, x_{j+k}; g) := \frac{\Delta(x_j, \ldots, x_{j+k-1}; g) - \Delta(x_{j+1}, \ldots, x_{j+k}; g)}{x_j - x_{j+k}}$$
$$\text{for } k = 1, \ldots, m \text{ and } j = \nu, \ldots, \nu + m - k. \tag{3.97}$$

In almost every textbook on numerical analysis (e.g. G.MEINARDUS & G.MERZ [1] or J.STOER & R.BULIRSCH [1]), one can find the proofs of the following elementary properties of the operator Δ:

Lemma 3.31: *a) The divided difference $\Delta(x_\nu, \ldots, x_{\nu+m}; g)$ can be written in the form*

$$\Delta(x_\nu, \ldots, x_{\nu+m}; g) = \sum_{\mu=\nu}^{\nu+m} \frac{g(x_\mu)}{\omega_{\nu,\mu}^m} \tag{3.98a}$$

with

$$\omega_{\nu,\mu}^m = \prod_{\substack{\lambda=\nu \\ \lambda\neq\mu}}^{\nu+m} (x_\mu - x_\lambda). \tag{3.98b}$$

b) Application of the divided difference operator Δ *to the monomial* $p_j(x) := x^j$ *yields*

$$\Delta(x_\nu, \ldots, x_{\nu+m}; p_j) = \begin{cases} 0, & \text{for } j = 0, \ldots, m-1 \\ 1, & \text{for } j = m \,, \end{cases} \tag{3.99}$$

i.e. Δ *annihilates the polynomial space* Π_{m-1}. *(In fact,* Δ *is even uniquely determined by (3.99), see* C.BREZINSKI & G.WALZ [1]*).*

Let $[x_\nu, \ldots, x_{\nu+m}]$ denote the smallest interval containing the points $x_\nu, \ldots, x_{\nu+m}$, and consider some $g \in C^m[x_\nu, \ldots, x_{\nu+m}]$. Then it is known that $\Delta(x_\nu, \ldots, x_{\nu+m}; g)$ converges to $\frac{g^{(m)}(\xi)}{m!}$, if all points $x_\nu, \ldots, x_{\nu+m}$ collapse to one, say ξ. In the case of equidistant points, the following fundamental lemma sharpens this result considerably:

Lemma 3.32 (G.WALZ [13]): *Fix some arbitrary index* $k \in \{\nu, \ldots, \nu+m\}$, *and consider, for* $n \in \mathbb{N}$, *the sequence of equidistant points* $\{x_\nu^{(n)}, \ldots, x_{\nu+m}^{(n)}\}$, *defined through*

$$x_{k+j}^{(n)} := x_k + \frac{j}{n}, \quad j = \nu - k, \ldots, \nu - k + m \,.$$

(Note that $x_k^{(n)} \equiv x_k$ *for all* n*).*
Furthermore, assume that the function g *belongs to the differentiability class* $C^{r+1}[x_\nu^{(1)}, \ldots, x_{\nu+m}^{(1)}]$ *for some* $r > m$.

Then the sequence $\{\Delta(x_\nu^{(n)}, \ldots, x_{\nu+m}^{(n)}; g)\}_{n\in\mathbb{N}}$ *possesses the following logarithmic asymptotic expansion of order* $r - m$:

$$\Delta(x_\nu^{(n)}, \ldots, x_{\nu+m}^{(n)}; g) =$$

$$= \frac{g^{(m)}(x_k)}{m!} + \sum_{j=1}^{r-m} \frac{c_j}{n^j} + O(n^{m-r-1}). \tag{3.100}$$

Proof. Since g is in $C^{r+1}[x_\nu^{(1)},\ldots,x_{\nu+m}^{(1)}]$, Taylor's formula yields for any $x_\mu^{(n)}$:

$$g(x_\mu^{(n)}) = \sum_{j=0}^{r} \frac{g^{(j)}(x_k)}{j!}(x_\mu^{(n)} - x_k)^j + O((x_\mu^{(n)} - x_k)^{r+1}). \quad (3.101)$$

A combination of this result with the representation (3.98) gives us the relation

$$\Delta(x_\nu^{(n)},\ldots,x_{\nu+m}^{(n)};g) = \quad (3.102)$$

$$= \sum_{\mu=\nu}^{\nu+m} \frac{1}{\omega_{\nu,\mu}^{m}} \left(\sum_{j=0}^{r} \frac{g^{(j)}(x_k)}{j!}(x_\mu^{(n)} - x_k)^j + O((x_\mu^{(n)} - x_k)^{r+1}) \right).$$

Now, since $x_\mu^{(n)} - x_k = (\mu - k)/n$, we obtain from (3.102)

$$\Delta(x_\nu^{(n)},\ldots,x_{\nu+m}^{(n)};g) =$$

$$= \sum_{\mu=\nu}^{\nu+m} \frac{n^m}{\prod\limits_{\substack{\lambda=\nu \\ \lambda\neq\mu}}^{\nu+m} (\mu-\lambda)} \sum_{j=0}^{r} \frac{g^{(j)}(x_k)}{j!}\frac{(\mu-k)^j}{n^j} + O(n^{m-r-1}) \quad (3.103)$$

$$= n^m \sum_{j=0}^{r} \frac{g^{(j)}(x_k)}{j!n^j} \sum_{\mu=\nu}^{\nu+m} \frac{(\mu-k)^j}{\prod\limits_{\substack{\lambda=\nu \\ \lambda\neq\mu}}^{\nu+m} (\mu-\lambda)} + O(n^{m-r-1}).$$

At this point of our investigation, it is important to note that

$$\sum_{\mu=\nu}^{\nu+m} \frac{(\mu-k)^j}{\prod\limits_{\substack{\lambda=\nu \\ \lambda\neq\mu}}^{\nu+m} (\mu-\lambda)} = \Delta(\nu,\ldots,\nu+m;(\,\cdot\,-k)^j)$$

$$= \begin{cases} 0, & \text{for } j = 0,\ldots,m-1 \\ 1, & \text{for } j = m\ , \end{cases}$$

due to (3.99). Inserting this into the last line of equation (3.103) finally proves the assertion. □

Corollary 3.33: *For a polynomial* $p \in \Pi_m$ *, we have*

$$\Delta(x_\nu, \ldots, x_{\nu+m}; p) = \frac{p^{(m)}(\xi)}{m!} \tag{3.104}$$

for each $\xi \in \mathbb{R}$ *.*

Remark. In the next subsection we will present a result on generalized divided differences, which almost covers Lemma 3.32 as a special case, but with a completely different proof. However, this "extra" proof of the lemma should be given here, since it is shorter and surely more elegant.

The result of Lemma 3.32 has a variety of applications in numerical analysis, the most obvious of which being the numerical differentiation of functions. This will be treated as an example in the rest of this section.

We still have the possibility to choose the index k appropriately, and the most promising (and popular) choices are $k = \nu$ and, for even values of m, $k = \nu + \frac{m}{2}$, which lead to the so-called forward resp. central difference operators (cf. also Subsection 3.1.3):

Definition 3.34: Fix some value $x \in \mathbb{R}$. Then, adopting the notations from above, we call the operator

$$\delta_n^f(g; m, x) := \Delta(x, x + \tfrac{1}{n}, \ldots, x + \tfrac{m}{n}; g) \tag{3.105}$$

the mth *forward difference* of g, and

$$\delta_n^c(g; m, x) := \Delta(x - \tfrac{m}{2n}, x - \tfrac{m-2}{2n}, \ldots, x + \tfrac{m-2}{2n}, x + \tfrac{m}{2n}; g) \tag{3.106}$$

the mth *central difference* of g. Note that we have on the right hand sides of (3.105) and (3.106) both times divided differences of order m.

We can now apply Lemma 3.32 to the operators $\delta_n^f(g;m,x)$ and $\delta_n^c(g;m,x)$, thus obtaining the main result of this subsection concerning the numerical differentiation of a function; in a different approach, i.e. without using divided differences, and with a completely different proof, the second part of the following result (concerning $\delta_n^c(g;m,x)$) goes back to H.RUTISHAUSER [1] (for $m = 1$) and T.STRÖM [3], see also H.ENGELS [1], T.STRÖM & J.N.LYNESS [1] and G.WALZ [1,7].

Theorem 3.35 (G.WALZ [13]): *Adopt the notations and assumptions from Lemma 3.32. Then the following results hold:*
a) The forward difference operator $\delta_n^f(g;m,x)$ possesses a logarithmic asymptotic expansion of the form

$$\delta_n^f(g;m,x) = \frac{g^{(m)}(x)}{m!} + \sum_{j=1}^{r-m} \frac{c_j}{n^j} + O(n^{m-r-1}). \tag{3.107}$$

b) The central difference operator $\delta_n^c(g;m,x)$ possesses a logarithmic asymptotic expansion of the form

$$\delta_n^c(g;m,x) = \frac{g^{(m)}(x)}{m!} + \sum_{j=1}^{r-m} \frac{\gamma_j}{n^j} + O(n^{m-r-1}) \tag{3.108}$$

with

$$\gamma_j = 0 \quad \text{for } j \text{ odd},$$

i.e. the expansion in (3.108) contains only even powers of n.

Proof. The assertion in a) follows directly from Lemma 3.32. Also, representation (3.108) is an immediate consequence of this lemma, and we only have to show that the $\gamma_j's$ with odd index vanish. But this follows from the fact that the central difference operator $\delta_n^c(g;m,x)$ is symmetric, i.e. we have

$$\delta_n^c(g;m,x) = \delta_{-n}^c(g;m,x) \tag{3.109}$$

for all n, and therefore the right hand side of (3.108) cannot contain odd powers of n. □

Although generalizations of divided differences to the multivariate setting are treated at several places in the literature (mainly in connection with multivariate interpolation or multivariate B-splines), their limiting behavior for collapsing nodes, in particular the existence of asymptotic expansions in this case is not investigated so far. However, the reader who is interested in multivariate divided differences (maybe in order to attack the above-mentioned problem) is referred to A.S.Cavaretta, C.A.Micchelli & A.Sharma [1], M.Gasca & A.López-Carmona [1], M.Gasca & J.I.Maeztu [1], H.A.Hakopian [1], C.A.Micchelli [1,2], M.Neamtu [1,2] and H.E.Salzer [2].

3.5.2 Generalized Divided Differences. In this subsection we want to work out a result corresponding to that of Lemma 3.32 for the case of generalized divided differences with respect to a so-called Polya space. Using the notation

$$D\begin{pmatrix} f_0 & f_1 & \cdots & f_m \\ y_0 & \cdots & \cdots & y_m \end{pmatrix} := \det \begin{pmatrix} f_0(y_0) & \cdots & \cdots & f_0(y_m) \\ \vdots & \vdots & \vdots & \vdots \\ \vdots & \vdots & \vdots & \vdots \\ f_m(y_0) & \cdots & \cdots & f_m(y_m) \end{pmatrix},$$

for functions $f_0, \ldots, f_m$ and numbers $y_0, \ldots, y_m$, we can give the following definition of generalized divided differences, which goes back to T.Popoviciu [1]:

Definition 3.36: Let there be given a set of functions $F := \{f_0, \ldots, f_m\}$ and $m+1$ pairwise distinct points $x_\nu, \ldots x_{\nu+m}$, such that

$$D\begin{pmatrix} f_0 & \cdots & f_{m-1} & f_m \\ x_\nu & \cdots & \cdots & x_{\nu+m} \end{pmatrix} \neq 0 \tag{3.110}$$

Then the *generalized divided difference* Δ_F of a function g with respect to the set F is defined as

$$\Delta_F(x_\nu, \ldots, x_{\nu+m}; g) := \frac{D\begin{pmatrix} f_0 & \cdots & f_{m-1} & g \\ x_\nu & \cdots & \cdots & x_{\nu+m} \end{pmatrix}}{D\begin{pmatrix} f_0 & \cdots & f_{m-1} & f_m \\ x_\nu & \cdots & \cdots & x_{\nu+m} \end{pmatrix}}. \tag{3.111}$$

Again, the number m is called the order of the divided difference.

Obviously Δ_F annihilates the linear space $\text{span}\{f_0, \ldots, f_{m-1}\}$ and is normalized by $\Delta_F(x_\nu, \ldots, x_{\nu+m}; f_m) = 1$. Further properties of the generalized divided differences can be found in many publications, e.g. S.Karlin [1], G.Mühlbach [1,3] and G.Walz [12,13]. The most important one for practical computations is surely the fact that the generalized divided differences can also be computed via a reccurence relation, very similar to that of the classical ones given in (3.97):

Theorem 3.37 (G.Mühlbach [1]): *Assume that the set of functions F forms a so-called complete Chebyshev system on the point set $\{x_\nu, \ldots x_{\nu+m}\}$, i.e. it is*

$$D\begin{pmatrix} f_0 & \cdots & \cdots & f_k \\ y_0 & \cdots & \cdots & y_k \end{pmatrix} > 0$$

for each $k \in \{0, \ldots, m\}$ and each subset $\{y_0 < \cdots < y_k\} \subset \{x_\nu, \ldots x_{\nu+m}\}$.

Then the generalized divided difference $\Delta_F(x_\nu, \ldots, x_{\nu+m}; g)$ can be computed via the following recurrence relation:

$$\Delta_F(x_j; g) = \frac{g(x_j)}{f_0(x_j)} \qquad \textit{for } j = \nu, \ldots, \nu+m,$$

$$\Delta_F(x_j, \ldots, x_{j+k}; g) = \tag{3.112}$$

$$= \frac{\Delta_F(x_j, \ldots, x_{j+k-1}; g) - \Delta_F(x_{j+1}, \ldots, x_{j+k}; g)}{\Delta_F(x_j, \ldots, x_{j+k-1}; f_k) - \Delta_F(x_{j+1}, \ldots, x_{j+k}; f_k)}$$

$$\textit{for } k = 1, \ldots, m \textit{ and } j = \nu, \ldots, \nu+m-k.$$

Proof. Denote for a moment by $\tilde{\Delta}$ that functional, which is defined through the recursion (3.112). Using induction it is easy to show that $\tilde{\Delta}$ is a linear functional of the form

$$\tilde{\Delta}_F(x_j, \dots, x_{j+k}; g) = \sum_{i=j}^{j+k} \alpha_{i,j} g(x_i) \,,$$

satisfying

$$\tilde{\Delta}_F(x_j, \dots, x_{j+k}; f_\mu) = \begin{cases} 0\,, & \text{for } \mu = 0, \dots, k-1 \\ 1\,, & \text{for } \mu = k \,. \end{cases}$$

But the same is true for the functional Δ itself, and so it follows from C.BREZINSKI & G.WALZ [1] that $\Delta = \tilde{\Delta}$. □

Remark. The assumption that F forms a complete Chebyshev system can be weakened (see G.MÜHLBACH [1]), but it is the appropriate one for the following considerations.

We turn now to the question, how the result of Lemma 3.32 can be generalized to the operator Δ_F and, simultaneously, what is a suitable definition for $\Delta_F(x_\nu, \dots, x_{\nu+m}; g)$, if all the points x_j collapse to a single one.

Of course this cannot be answered without any further information on the set of functions F, and it turns out that the following reasonable generalization of the polynomials is the right one for our purposes:

Definition 3.38: Let there be given $m+1$ functions $w_0, \dots, w_m$, which are sufficiently smooth and strictly positive on some given interval $[a, b]$.

We define the functions $v_0, \ldots, v_m$ by

$$\begin{aligned}
v_0(x) &= w_0(x) \\
v_1(x) &= w_0(x) \int_a^x w_1(t_1) dt_1 \\
v_2(x) &= w_0(x) \int_a^x w_1(t_1) \int_a^{t_1} w_2(t_2)\, dt_2\, dt_1 \\
&\vdots \\
v_m(x) &= w_0(x) \int_a^x w_1(t_1) \int_a^{t_1} w_2(t_2) \cdots \\
&\qquad \cdots \int_a^{t_{m-1}} w_m(t_m)\, dt_m\, dt_{m-1} \cdots dt_1
\end{aligned} \tag{3.113}$$

and denote the set of functions $V := \{v_0, \ldots, v_m\}$ as *Polya system*; moreover, the linear space spanned by these function will be denoted as *Polya space*, see also S.KARLIN & W.J.STUDDEN [1] and G.MEINARDUS & G.WALZ [2].

It is known (see S.KARLIN & W.J.STUDDEN [1]) that V forms a complete Chebyshev system on $[a, b]$. Standard examples for Polya systems are the polynomials, which are recovered by setting $w_0(x) := 1$ and $w_j(x) := j$ for $j = 1, \ldots, m$, and the exponential sums $\{\exp(\lambda_0 x), \ldots, \exp(\lambda_m x)\}$, which we obtain by choosing

$$\begin{aligned}
w_0(x) &:= \exp(\lambda_0 x) \quad \text{and} \\
w_j(x) &:= \exp((\lambda_j - \lambda_{j-1})x) \quad \text{for } j = 1, \ldots, m.
\end{aligned}$$

Now, let us come back to the general case and consider the linear differential operator L_m of order m, defined through

$$\begin{aligned}
L_0(g)(x) &:= \frac{1}{w_0(x)} \cdot g(x), \\
L_\mu(g)(x) &:= \frac{1}{w_\mu(x)} \cdot \frac{d}{dx} L_{\mu-1}(g)(x) \quad \text{for } \mu = 1, \ldots, m,
\end{aligned} \tag{3.114}$$

with $g \in C^m[a,b]$ and the functions w_μ from above. Obviously, the operator L_m is normalized such that

$$L_m(v_j) \;=\; \begin{cases} 0\,, & \text{for } j = 0,\ldots,m-1 \\ 1\,, & \text{for } j = m\,, \end{cases} \tag{3.115}$$

We now prove the main result of this section, which yields an asymptotic expansion for the sequence of generalized divided differences Δ_F , if all the points x_j converge to one. To avoid unnecessary notational difficulties (we shall have sub- and superscripts enough to struggle with) we assume that these points converge to the *first* one (i.e., in the language of Lemma 3.32, we take $k = \nu$), and that all functions under consideration belong to the class $C^\infty[a,b]$. Then the following result holds.

Theorem 3.39: *Choose some real point x_ν and consider, for $n \in I\!N$, the sequence of equidistant points $\{x_\nu^{(n)},\ldots,x_{\nu+m}^{(n)}\}$, defined through*

$$x_{\nu+j}^{(n)} \;:=\; x_\nu + \frac{j}{n}\,, \quad j = 0,\ldots,m\,.$$

Furthermore, assume that g belongs to the differentiability class $C^\infty[x_\nu^{(1)},\ldots,x_{\nu+m}^{(1)}]$.

Then the sequence of generalized divided differences

$$\{\Delta_V(x_\nu^{(n)},\ldots,x_{\nu+m}^{(n)};g)\}_{n\in I\!N}\,,$$

taken with respect to the Polya space V , possesses an asymptotic expansion of the form

$$\Delta_V(x_\nu^{(n)},\ldots,x_{\nu+m}^{(n)};g) \;=\; L_m(g)(x_\nu) \;+\; \sum_{\mu=1} \frac{c_\mu^{(m)}(x_\nu)}{n^\mu} \tag{3.116}$$

with coefficient functions $c_\mu^{(m)} = c_\mu^{(m)}(V,g)$. Here, L_m denotes the linear differential operator defined in (3.114).

Proof. We prove the assertion by induction with respect to m, starting with $m = 1$. In this case, the generalized divided difference reads explicitly

$$\Delta_V(x_\nu, x_{\nu+1}; g) = \frac{v_0(x_\nu)\cdot g(x_{\nu+1}) - v_0(x_{\nu+1})\cdot g(x_\nu)}{v_0(x_\nu)\cdot v_1(x_{\nu+1}) - v_0(x_{\nu+1})\cdot v_1(x_\nu)}. \quad (3.117)$$

(From now on we shall drop the upper index "(n)" of the x_j's). Let us treat the numerator of (3.117) first; if we expand the functions v_0 and g in a formal Taylor series about x_ν and observe that $x_{\nu+1} - x_\nu = 1/n$, we see that the numerator is of the form

$$\begin{aligned} v_0(x_\nu)\cdot g(x_{\nu+1}) - v_0(x_{\nu+1})\cdot g(x_\nu) \; &= \\ = \; v_0(x_\nu)\cdot \sum_{j=0}^{\infty} \frac{g^{(j)}(x_\nu)}{j!\, n^j} \; &- \; g(x_\nu)\cdot \sum_{j=0}^{\infty} \frac{v_0^{(j)}(x_\nu)}{j!\, n^j} \\ = \; \frac{v_0(x_\nu)\cdot g'(x_\nu) - v_0'(x_\nu)\cdot g(x_\nu)}{n} \; &+ \; \sum_{j=2}^{\infty} \frac{\gamma_j(x_\nu)}{n^j} \quad (3.118) \end{aligned}$$

with certain coefficient functions γ_j, which depend on g and v_0, but not on n.

A similar technique applies to the denominator of (3.117); first we note that

$$\begin{aligned} v_0(x_\nu)\cdot v_1(x_{\nu+1}) - v_0(x_{\nu+1})\cdot v_1(x_\nu) \; &= \\ = \; w_0(x_\nu)\cdot w_0(x_{\nu+1})\cdot \int_a^{x_{\nu+1}} w_1(t_1)\, dt_1 & \\ - \; w_0(x_{\nu+1})\cdot w_0(x_\nu)\cdot \int_a^{x_\nu} w_1(t_1)\, dt_1 & \\ = \; w_0(x_\nu)\cdot w_0(x_{\nu+1})\cdot \int_{x_\nu}^{x_{\nu+1}} w_1(t_1)\, dt_1\,, \quad (3.119) & \end{aligned}$$

and expanding the functions w_0 and w_1 about x_ν yields – after integration – that the expression in (3.119) can be written in the form

$$\begin{aligned} w_0(x_\nu)\cdot\sum_{j=0}^{\infty}\frac{w_0^{(j)}(x_\nu)}{j!\,n^j}\cdot\sum_{\mu=0}^{\infty}\frac{w_1^{(\mu)}(x_\nu)}{(\mu+1)!\,n^{(\mu+1)}} &= \\ = \frac{w_0(x_\nu)^2\cdot w_1(x_\nu)}{n} + \sum_{\kappa=2}^{\infty}\frac{\delta_\kappa(x_\nu)}{n^\kappa}, \end{aligned} \tag{3.120}$$

where again the coefficient functions δ_κ do not depend on n. Now, a combination of equations (3.117), (3.118) and (3.120) leads us to the intermediate result

$$\Delta_V(x_\nu, x_{\nu+1}; g) = \frac{w_0(x_\nu)\cdot g'(x_\nu) - w_0'(x_\nu)\cdot g(x_\nu)}{w_0(x_\nu)^2\cdot w_1(x_\nu)} + \sum_{\mu=1} \frac{c_\mu^{(1)}(x_\nu)}{n^\mu} \tag{3.121}$$

(note that $w_0 = v_0$). On the other hand, the explicit form of $L_1(g)(x_\nu)$ is, following (3.114),

$$L_1(g)(x_\nu) = \frac{1}{w_1(x_\nu)}\cdot\frac{w_0(x_\nu)\cdot g'(x_\nu) - w_0'(x_\nu)\cdot g(x_\nu)}{w_0(x_\nu)^2} \tag{3.122}$$

which proves, in combination with equation (3.121), the assertion for $m = 1$.

Now we assume, for some $m \geq 2$, that the assertion holds for $m-1$, and have to show that this is also the case for m itself. According to Theorem 3.37, the mth order generalized divided difference $\Delta_V(x_\nu, \ldots, x_{\nu+m}; g)$ is equal to

$$\frac{\Delta_V(x_\nu, \ldots, x_{\nu+m-1}; g) - \Delta_V(x_{\nu+1}, \ldots, x_{\nu+m}; g)}{\Delta_V(x_\nu, \ldots, x_{\nu+m-1}; v_m) - \Delta_V(x_{\nu+1}, \ldots, x_{\nu+m}; v_m)}\,. \tag{3.123}$$

Each of the four generalized divided differences on the right hand side of (3.123) is of order $m-1$, so we can apply the induction hypothesis. Therefore, it is clear that the expression in (3.123) possesses an asymptotic expansion of the desired type, and all we are left to do is to show that the limit is the one we want it to be.

Let us treat the numerator first. According to the hypothesis, we may write

$$\Delta_V(x_\nu,\ldots,x_{\nu+m-1};g) = L_{m-1}(g)(x_\nu) + \sum_{\mu=1} \frac{c_\mu^{(m-1)}(x_\nu)}{n^\mu} \tag{3.124a}$$

and

$$\Delta_V(x_{\nu+1},\ldots,x_{\nu+m};g) = L_{m-1}(g)(x_{\nu+1}) + \sum_{\mu=1} \frac{c_\mu^{(m-1)}(x_{\nu+1})}{n^\mu}\,. \tag{3.124b}$$

Proceeding almost exactly as in the case $m = 1$, we expand the function $L_{m-1}(g)$ about x_ν and find that the difference of the two constant terms in the numerator in (3.123) can be written as

$$L_{m-1}(g)(x_\nu) - L_{m-1}(g)(x_{\nu+1}) = \sum_{j=1}^{\infty} \frac{\gamma_j(x_\nu)}{n^j} \tag{3.125}$$

with

$$\gamma_1 = -\frac{d}{dx} L_{m-1}(g)\,. \tag{3.126}$$

The same arguments hold for the coefficient functions $c_j^{(m-1)}$ in (3.124), in particular for $c_1^{(m-1)}$, yielding the representation

$$c_1^{(m-1)}(x_\nu) - c_1^{(m-1)}(x_{\nu+1}) = \sum_{j=1}^{\infty} \frac{\tilde{\gamma}_j(x_\nu)}{n^j} \tag{3.127}$$

with certain coefficients $\tilde{\gamma}_j$. Packing now together our results from (3.124) - (3.127), we see that the numerator in (3.123) possesses an asymptotic expansion of the form

$$\Delta_V(x_\nu,\ldots,x_{\nu+m-1};g) - \Delta_V(x_{\nu+1},\ldots,x_{\nu+m};g) = \sum_{j=1} \frac{\alpha_j(x_\nu)}{n^j} \tag{3.128a}$$

with

$$\alpha_1 = -\frac{d}{dx} L_{m-1} \,. \tag{3.128b}$$

(Note that is important to have the result (3.127), since otherwise we could not exclude that the coefficient $c_1^{(m-1)}$ from (3.124) has influence on α_1).

The treatment of the denominator in (3.123) is now easy, since we have all the tools together: Again according to the hypothesis, we may write

$$\Delta_V(x_\nu, \ldots, x_{\nu+m-1}; v_m) \;=\; L_{m-1}(v_m)(x_\nu) + \sum_{\mu=1} \frac{\tilde{c}^{(m-1)}(x_\nu)}{n^\mu}$$

and the same for $\Delta_V(x_{\nu+1}, \ldots, x_{\nu+m}; v_m)$. Now we observe that

$$L_{m-1}(v_m)(x) \;=\; \int_a^x w_m(t_m)\, dt_m \,,$$

hence (cf. (3.119) and (3.120))

$$\begin{aligned} L_{m-1}(v_m)(x_\nu) - L_{m-1}(v_m)(x_{\nu+1}) \;&=\; -\int_{x_\nu}^{x_{\nu+1}} w_m(t_m)\, dt_m \\ &= \sum_{j=1}^{\infty} \frac{\delta_j(x_\nu)}{n^j} \quad \text{with} \quad \delta_1 = -w_m \,. \end{aligned} \tag{3.129}$$

Finally, we put together (3.123), (3.128) and (3.129) and obtain

$$\begin{aligned} &\Delta_V(x_\nu, \ldots, x_{\nu+m}; g) \;= \\ &\quad = \frac{1}{w_m(x_\nu)} \cdot \frac{d}{dx} L_{m-1}(g)(x_\nu) \;+\; \sum_{\mu=1} \frac{c_\mu^{(m)}(x_\nu)}{n^\mu} \\ &\quad = L_m(g)(x_\nu) \;+\; \sum_{\mu=1} \frac{c_\mu^{(m)}(x_\nu)}{n^\mu} \,, \end{aligned}$$

which completes the proof of Theorem 3.39. □

Obviously, our remarks from the end of Subsection 3.5.1 concerning the numerical differentiation of functions do also apply here (where now "differentiation" means: computation of the operator L_m), but we do not want to work this out further. Instead, we would like to point out that the result of Theorem 3.39 makes it possible to define generalized divided differences for multiple points in a consistent manner, cf. also G.MÜHLBACH [2] or G.WALZ [13]. For $\xi \in I\!R$, it is

$$\Delta_V(\underbrace{\xi,\xi,\ldots,\xi}_{(m+1)\ times};g) := L_m(g)(\xi); \tag{3.130}$$

note that this definition is in agreement with the usual one for classical divided differences, i.e.

$$\Delta(\underbrace{\xi,\xi,\ldots,\xi}_{(m+1)\ times};g) = \frac{g^{(m)}(\xi)}{m!}.$$

3.5.3 The E-Method. The method we are going to study on the next pages is by no means restricted to numerical differentiation, but since differentiation is until now its main application, we have decided to present it in this section. The E-method is used in order to condense a logarithmic asymptotic expansion in a certain sense; to make this statement more precise, we have to go back for a moment to the "roots", i.e. to the definition of a logarithmic asymptotic expansion. At the very beginning of Chapter 3 we have learned that a sequence $\{\sigma_n\}$ possesses a logarithmic asymptotic expansion of order m, if the elements σ_n can be written in the form

$$\sigma_n = c_0 + \sum_{\mu=1}^{m} \frac{c_\mu}{n^{\rho_\mu}} + o(n^{-Re\rho_m}) \text{ for } n \to \infty . \tag{3.131}$$

Now we want to specialize to that subclass of logarithmic asymptotic expansions, where the exponents ρ_μ are all integer multiples of one positive number $\eta \in I\!R$; we have already seen that this is by far the most important case in the applications.

Definition 3.40: A logarithmic asymptotic expansion as in (3.131) will be denoted as *special logarithmic asymptotic expansion* (of order m), if there exists a positive real number η, such that each σ_n can be written in the form

$$\sigma_n = c_0 + \sum_{\mu=1}^{m} \frac{c_\mu}{n^{\eta\cdot\mu}} + o(n^{-\eta\cdot m}) \text{ for } n \to \infty . \tag{3.132}$$

If η is maximal, which means that there is no $\eta_1 > \eta$ with (3.132), then η is denoted as the *modulus* of this asymptotic expansion.

The aim of this subsection is to present a method for the construction of asymptotic expansions with *arbitrarily large* modulus (starting with a given one of modulus 1), the so-called E-method. We shall see later on (see Subsection 5.3.1) that increasing the modulus not only makes the extrapolation process, applied to the asymptotic expansion *faster* (and therefore more efficient), but also increases the numerical stability.

So, consider now a given sequence $\{\sigma_n\}$ of type (3.132) with $\eta = 1$ of arbitrary order, thus

$$\sigma_n = c_0 + \sum_{\mu=1} \frac{c_\mu}{n^\mu} . \tag{3.133}$$

Suppose furthermore that also the $\sigma_n' s$ with negative index, i.e. all elements of the sequence $\{\sigma_{-n}\}_{n\in I\!N}$ are well-defined. Then we may also consider the new sequence $\{\tilde{\sigma}_n\}$, defined for all n through

$$\tilde{\sigma}_n := \frac{1}{2}(\sigma_n + \sigma_{-n}) , \tag{3.134}$$

and it is obvious that this sequence possesses a special logarithmic asymptotic expansion with modulus 2. This construction is quite old, see for example L.F.RICHARDSON [3] or H.RUTISHAUSER [1], but now we shall work out that it can be generalized and improved considerably to the following mean–value method:

We denote, for some fixed $K \in I\!N$, by $\zeta = \zeta_K$ the primitive K^{th} root of unity:

$$\zeta := \exp(\frac{2\pi i}{K}) .$$

With the elements σ_n from (3.133) we define for $\kappa = 0, \ldots, K-1$ and all $n \in I\!N$

$$\sigma_n^{[\kappa]} := \sigma_{\zeta^\kappa \cdot n}, \tag{3.135}$$

where it is assumed that the values on the right–hand side are well–defined. Then it is easily seen that these $\sigma_n^{[\kappa]}$ also possess a special logarithmic asymptotic expansion with modulus 1 since

$$\sigma_n^{[\kappa]} = c_0 + \sum_{\mu=1} \frac{c_\mu}{\zeta^{\kappa\mu}} \cdot \frac{1}{n^\mu} := c_0 + \sum_{\mu=1} \frac{c_\mu^{[\kappa]}}{n^\mu} \tag{3.136}$$

with $|c_\mu^{[\kappa]}| = |c_\mu|$ for all $\mu = 1, \ldots, \infty$ and $\kappa = 0, \ldots, K-1$. We say that the $\sigma_n^{[\kappa]}$ are *based on* the Kth roots of unity. Using these auxiliary sequences, we now construct an expansion with modulus K:

Theorem 3.41 (G.WALZ [1,6,7]): *Let $\{\sigma_n\}$ be a sequence of the type (3.133) and assume that the $\sigma_n^{[\kappa]}$ from (3.135) are well-defined for all κ and n. Then the sequence $\{\Sigma_n^{[K]}\}_{n \in I\!N}$, defined through*

$$\Sigma_n^{[K]} := \frac{1}{K} \cdot \sum_{\kappa=0}^{K-1} \sigma_n^{[\kappa]} \tag{3.137}$$

for all $n \in I\!N$, possesses a special logarithmic asymptotic expansion with modulus K.

Proof. This is just an application of the properties of $\zeta = \zeta_K$, in particular the fact that

$$\sum_{\kappa=0}^{K-1} \frac{1}{\zeta^{\kappa\mu}} = \begin{cases} K & , \quad \text{if } K \mid \mu, \\ 0 & \quad \text{otherwise} \end{cases} \tag{3.138}$$

(cf. also Problem P 3.11).

We may execute the straightforward calculation

$$\begin{aligned} \Sigma_n^{[K]} &= \frac{1}{K} \cdot \sum_{\kappa=0}^{K-1} \sigma_n^{[\kappa]} \\ &= c_0 + \frac{1}{K} \sum_{\kappa=0}^{K-1} \sum_{\mu=1} \frac{c_\mu}{\zeta^{\kappa\mu}} \cdot \frac{1}{n^\mu} \\ &= c_0 + \frac{1}{K} \sum_{\mu=1} \frac{c_\mu}{n^\mu} \cdot \sum_{\kappa=0}^{K-1} \frac{1}{\zeta^{\kappa\mu}} \\ &\overset{(3.138)}{=} c_0 + \sum_{\lambda=1} \frac{c_{K\lambda}}{n^{K\lambda}} , \end{aligned}$$

which completes the proof of Theorem 3.41. □

Obviously, there are no restrictions concerning the integer K , so (3.137) defines a method for the construction of logarithmic asymptotic expansions with arbitrarily large modulus. Due to the strong connection of this method with the roots of unity (= "Einheitswurzeln" in German), it is denoted as *E-method*.

There is an interesting formal analogy of the E-method with the Fast Fourier Transform, a very prominent method in the theory of fast computations:

Suppose that K is even and define for $n \in I\!N$ and $\kappa = 0, \ldots, \frac{K}{2} - 1$:

$$\tilde{\sigma}_n^{[\kappa]} := \frac{1}{2}(\sigma_n^{[\kappa]} + \sigma_n^{[K/2+\kappa]}) .$$

A short computation yields that

$$\tilde{\sigma}_n^{[\kappa]} = c_0 + \sum_{\mu=1} \frac{c_{2\mu}}{n^{2\mu}} \cdot \frac{1}{\zeta^{2\kappa\nu}} ,$$

i.e. the K expansions $\{\sigma_n^{[0]}, \ldots, \sigma_n^{[K-1]}\}$, which had modulus 1 and were based on the Kth roots of unity, have been transformed into $\frac{K}{2}$ expansions with modulus 2, based on the $(K/2)^{th}$ roots of unity (note that $\zeta_K^2 = \zeta_{\frac{K}{2}}$).

Now, if $\frac{K}{2}$ again is an even number, we can proceed as above and obtain $\frac{K}{4}$ expansions with modulus 4, based on the $(K/4)^{th}$ roots of unity, and so on. Summarizing we get

Lemma 3.42: *With an arbitrary number $J \in \mathbb{N}$ let $K = 2^J$ and $\sigma_n^{[\kappa]}$ defined as in (3.135). Furthermore, define the sequences $\{\sigma_n^{\kappa,j}\}$ as follows:*

$$\begin{aligned} \sigma_n^{\kappa,0} &= \sigma_n^{[\kappa]}, && \kappa = 0, \ldots, K-1, \\ \sigma_n^{\kappa,j} &= \frac{1}{2}(\sigma_n^{\kappa,j-1} + \sigma_n^{K/2^j+\kappa,j-1}), && \begin{cases} j = 1, \ldots, J, \\ \kappa = 0, \ldots, K/2^j - 1. \end{cases} \end{aligned} \tag{3.139}$$

Then the sequences $\{\sigma_n^{\kappa,j}\}$, $\kappa = 0, \ldots, \frac{K}{2^j} - 1$, possess for $j = 0, \ldots, J$ a special logarithmic asymptotic expansion with modulus 2^j ; they are based on the $(K/2^j)^{th}$ roots of unity. In particular we have

$$\sigma_n^{0,J} \equiv \Sigma_n^{[K]}.$$

Note the analogy of (3.139) to the Fast Fourier Transform, in particular to the version given in G.MEINARDUS [2] (see also G.MEINARDUS & G.MERZ [1]): in both cases two terms, which are based on opposite sides of the unit circle, are combined recursively, and so the number of terms is in each step precisely halved, whereas their complexity (here modulus, there degree) is doubled.

Of course the applicability of the E-method is restricted to those cases, where the underlying sequences $\{\sigma_n\}$ are well-defined also for complex values of the index (or at least, if $K = 2$, for negative real ones). As a first example, we recall from Subsection 3.3.2 the construction of logarithmic asymptotic expansions with a prescribed limit function f. In Theorem 3.13 we had seen that the sequence $\{\sigma_n(x)\}$, defined through

$$\sigma_n(x) = n \cdot a_1 \cdot f^{-1}\left(\frac{f(x)}{n}\right)$$

possesses a logarithmic asymptotic expansion of arbitrary order with limit $f(x)$.

If f and f^{-1} can be continued into the complex plane (more precisely, onto the closed unit circle), the E-method can be applied. Moreover, if we are in the situation of Subsection 3.3.3, i.e. if the function f^{-1} satisfies an addition theorem, then the values of $\sigma_{\zeta^\kappa \cdot n}(x)$ for $\kappa = 0, \ldots, K-1$ can be computed for all $n = 2^i$ by means of *the same recurrence relation* as in the case $\zeta \equiv 1$ (Theorem 3.17). The only additional difficulty comes from the need of the starting values

$$\sigma_{\zeta^\kappa \cdot 1}(x) \;=\; \zeta^\kappa \cdot a_1 \cdot f^{-1}\left(f(x) \cdot \zeta^{-\kappa}\right) ,$$

which have to be computed for $\kappa = 0, \ldots, K-1$.

In order to illustrate the efficiency of the E-method, we will now apply it in some detail to the asymptotic expansions worked out in Subsection 3.5.1.

Let g be an entire function and z an arbitrary fixed complex number. Suppose that we want to approximate the first derivative of g in z, $g'(z)$. To accomplish this, we make use of the forward difference operator δ_n^f as defined in (3.105), i.e.

$$\delta_n^f(g; 1, z) \;=\; n \cdot (g(z + \frac{1}{n}) - g(z)) , \tag{3.140}$$

which possesses an special logarithmic asymptotic expansion of arbitray order with modulus 1. Therefore we may apply the E-method and obtain the following result:

Lemma 3.43: *Let ζ denote the primitive K^{th} root of unity and define, in analogy to (3.137), the difference operator $\Delta_n^{[K],f}$ by*

$$\Delta_n^{[K],f}(g; 1, z) \;:=\; \frac{1}{K} \cdot \sum_{\kappa=0}^{K-1} \delta_{\zeta^\kappa \cdot n}^f(g; 1, z) ; \tag{3.141}$$

then $\Delta_n^{[K],f}$ *possesses a special logarithmic asymptotic expansion with limit* $g'(z)$ *and modulus* K *:*

$$\Delta_n^{[K],f}(g;1,z) \;=\; g'(z) \;+\; \sum_{\lambda=1} \frac{c_{K\lambda}(z)}{n^{K\lambda}} \,. \tag{3.142}$$

There is a nice and instructive geometric interpretation of this situation: The "simple" operator $\delta_n^f(g;1,z)$ from (3.140) approximates $g'(z)$ by evaluating the behavior of $g(w)$, as w approaches z "from the right". But g is an entire function, so we could as well let w approach z from other sides, and, using the resulting values of $g(w)$, we can expect to end up with a better approximation of $g'(z)$.

This is precisely what the E-method does, as can be seen from the following explicit form of (3.141):

$$\Delta_n^{[K],f}(g;1,z) \;=\; \frac{1}{K} \cdot \sum_{\kappa=0}^{K-1} \zeta^\kappa\, n \cdot (g(z+\frac{1}{\zeta^\kappa \cdot n}) - g(z)) \,. \tag{3.143}$$

This means that $\Delta_n^{[K],f}(g;1,z)$ evaluates the function g on K equidistant points on a circle around z with radius $\frac{1}{n}$, i.e. it collects information on g from K different directions of approach, and it is no wonder that this gives a better approximation on the derivative $g'(z)$.

Remark. Using the integration path $\gamma_n(t) := z + \frac{1}{n}e^{-it}$, $n \in \mathbb{N}$, $0 \le t \le 2\pi$, Cauchy's integral formula for the first derivative of g reads

$$g'(z) \;=\; \frac{n}{2\pi} \int_0^{2\pi} \mathrm{e}^{it} \cdot g(z+\frac{1}{\mathrm{e}^{it}\cdot n})\, dt \,.$$

This is exactly the continuous analogue of $\Delta_n^{[K],f}(g;1,z)$, since (3.143) can be rewritten once again as

$$\Delta_n^{[K],f}(g;1,z) \;=\; \frac{n}{K} \cdot \sum_{\kappa=0}^{K-1} \zeta^\kappa \cdot g(z+\frac{1}{\zeta^\kappa \cdot n}),$$

due to (3.138). Therefore, $\Delta_n^{[K],f}(g;1,z)$ could be denoted as *discrete Cauchy integral formula.*

We come now to the main result concerning the application of the E-method to the mth order difference operators from Subsection 3.5.1.

Theorem 3.44 (G.WALZ [1,6,7]): *Let g be an entire function, z an arbitrary complex number and K and m natural numbers; furthermore, let $\zeta = \zeta_K$ again denote the Kth root of unity. Define for $n \in \mathbb{N}$ the following finite-difference operators:*

$$\Delta_n^{[K],f}(g;m,z) := \frac{n^m}{K}\sum_{\nu=0}^{m}(-1)^{m-\nu}\binom{m}{\nu}\sum_{\kappa=0}^{K-1}\zeta^{\kappa m}g\Big(z+\frac{\nu}{\zeta^{\kappa}n}\Big) \tag{3.144}$$

and

$$\Delta_n^{[K],c}(g;m,z) := \frac{n^m}{K}\sum_{\nu=0}^{m}(-1)^{m-\nu}\binom{m}{\nu}\sum_{\kappa=0}^{K-1}\zeta^{\kappa m}g\Big(z+\frac{2\nu-m}{2\zeta^{\kappa}n}\Big) \tag{3.145}$$

Then both $\Delta_n^{[K],f}(g;m,z)$ and $\Delta_n^{[K],c}(g;m,z)$ possess a special logarithmic asymptotic expansion. It is of the form

$$\Delta_n^{[K],f}(g;m,z) = \frac{g^{(m)}(z)}{m!} + \sum_{\mu=1}\frac{c_\nu(z)}{n^{K\mu}} \tag{3.146}$$

and

$$\Delta_n^{[K],c}(g;m,z) = \frac{g^{(m)}(z)}{m!} + \sum_{\mu=1}\frac{\tilde{c}_\mu(z)}{n^{\eta\mu}} \tag{3.147}$$

respectively. In the second case, the modulus η equals

$$\eta = \begin{cases} K\,, & \text{if } K \text{ is even,} \\ 2K\,, & \text{if } K \text{ is odd.} \end{cases}$$

Proof. Let us treat the operator $\Delta_n^{[K],f}(g;m,z)$ first. It is an easy exercise to prove that the forward difference operator $\delta_n^f(g;m,z)$ has the explicit representation

$$\delta_n^f(g;m,z) = \frac{n^m}{m!}\sum_{\nu=0}^{m}(-1)^{m-\nu}\binom{m}{\nu}g\big(z+\frac{\nu}{n}\big).$$

Furthermore, we know from Theorem 3.35 that it possesses a special logarithmic asymptotic expansion with modulus 1 and limit $g^{(m)}/m!$. Therefore Theorem 3.41 yields that

$$\frac{1}{K}\cdot\sum_{\kappa=0}^{K-1}\delta_{\zeta^\kappa\cdot n}^f(g;m,z) =$$

$$= \frac{1}{K}\sum_{\kappa=0}^{K-1}\frac{\zeta^{\kappa m}n^m}{m!}\cdot\sum_{\nu=0}^{m}(-1)^{m-\nu}\binom{m}{\nu}g\big(z+\frac{\nu}{\zeta^\kappa\cdot n}\big) \quad (3.148)$$

possesses an asymptotic expansion of the form

$$\frac{g^{(m)}(z)}{m!} + \sum_{\mu=1}\frac{d_\mu(z)}{n^{K\mu}},$$

and re-ordering the terms in (3.148) finally proves (3.146).

The proof of (3.147) runs the same lines as the one we have just examined. To prove the fact that for odd values of K the modulus even equals $2K$, we notice that already $\delta_n^c(g;m,z)$ possesses an asymptotic expansion with modulus 2; now, taking into account that for odd K

$$\sum_{\kappa=0}^{K-1}\frac{1}{\zeta^{2\kappa\mu}} = \begin{cases} K, & \text{if } K \mid 2\mu \text{, i.e. if } K \mid \mu, \\ 0 & \text{otherwise}, \end{cases}$$

the assertion follows exactly as in the proof of Theorem 3.41. □

We close this section with a little numerical *example*: Computation of the second derivative of a function g in a point $z \in \mathbb{C}$.
a) An approximative solution of this problem is given by

$$\Delta_n^{[4],c}(g;2,z) = \frac{n^2}{2}\left(g(z+\frac{1}{n}) + g(z-\frac{1}{n}) - g(z-\frac{i}{n}) - g(z+\frac{i}{n})\right)$$

$$= g^{(2)}(z) + \sum_{\lambda=1} \frac{\gamma_\lambda(z)}{n^{4\lambda}} . \qquad (3.149)$$

Note that the computation of (3.149) essentially requires 4 evaluations of the function g; with this input only, it yields an approximating sequence whose error tends to 0 with n^{-4} and, what is even more important regarding extrapolation, possesses a logarithmic asymptotic expansion with modulus 4.

For the function $g(z) = \sin(z)$ and the point $z = (1+i)\frac{\pi}{2}$, where i denotes the complex unity, Table 3.3 shows the errors e_n of $\Delta_n^{[4],c}(g;2,z)$ compared to the true value $g^{(2)}(z) = -2.50917847\ldots$ for $n = 2^j$, $j = 0,\ldots,4$, and the quotients $e_n/e_{\frac{n}{2}}$ of two subsequent errors. Obviously, these quotients tend to 2^4, as predicted by (3.149).

n	Error of $\Delta_n^{[4],c}(g;2,z)$	Quotients of Errors
1	0.69713 $e(-2)$	
2	0.43562 $e(-3)$	16.00298
4	0.27226 $e(-4)$	16.00019
8	0.17016 $e(-5)$	16.00001
16	0.10635 $e(-6)$	16.00000

Table 3.3

b) Even higher is the efficiency of the operator $\Delta_n^{[3],c}(g;2,z)$, since, according to Theorem 3.44, it has an asymptotic expansion with modulus 6. Table 3.4 shows, in analogy to Table 3.3, the errors of $\Delta_n^{[3],c}(g;2,z)$ and the related quotients.

n	Error of $\Delta_n^{[3],c}(g;2,z)$	Quotients of Errors
1	0.12446 $e(-3)$	
2	0.19447 $e(-5)$	63.99814
4	0.30386 $e(-7)$	63.99997
8	0.47478 $e(-9)$	64.00000

Table 3.4

3.6 Discretization Methods for Ordinary Differential Equations

Discretization methods for the numerical solution of differential equations establish, apart from the problem of computing the value of the number π, the oldest discipline in numerical analysis, where asymptotic expansions and extrapolation methods play a prominent role (cf. also our Historical Notes). The basic idea (existence of an asymptotic expansion allows convergence acceleration by linear combination of two sequence elements) appeared in this context the first time in Richardson's paper from 1910 (L.F.RICHARDSON [1]), which was mainly concerned with partial differential equations, and was explained in more detail by Richardson and Gaunt in their famous quasi-joint paper of 1927 (L.F.RICHARDSON [3], J.A.GAUNT [1]), see also Subsection 4.3.2. However, even in this paper the analytical foundations of the method, in particular the proof of the asymptotic expansion, appeared somehow diffuse, and therefore extrapolation "remained a technique with an unclear domain of applicability" (P.DEUFLHARD [2]).

As far as differential equations are concerned, the great breakthrough came some 40 years later with W. Gragg's famous thesis W.B.GRAGG [1], (see also W.B.GRAGG [2]), written under the advice of P. Henrici, where the existence of an asymptotic expansion for several discretization methods for ordinary differential equations was finally proved (see e.g. Theorem 3.46 below).

Since then, the interest in these methods has grown exponentially; it is by far not possible to consider here all the different types of differential problems, where asymptotic expansions are involved, neither to present a complete list of the hundreds of references on this topic. Instead, we concentrate on those problems resp. methods which appear to be the most important on one hand, and instructive on the other in that sense that the basic ideas come out very clearly; we will treat therefore on the next pages mainly one-step and (linear) multistep methods for first order initial value problems. Some other approaches (Störmer's method for a second order initial value problem, stiff differential equations and differential-algebraic problems) will be looked at in a final short subsection. In addition, we would like to refer the reader to the following books and survey papers, where some additional information on specific problems can be found: P.DEUFLHARD [2], E.HAIRER, S.P.NØRSETT & G.WANNER [1], E.HAIRER & G.WANNER [1], D.C.JOYCE [1], G.I.MARCHUK & V.V.SHAIDUROV [1] and H.J. STETTER [3].

Moreover, we assume the reader to be familiar with the (very) basic theory of ordinary differential equations, in particular with methods for their numerical solution (if not, have a look in any textbook on the topic, e.g. C.W.GEAR [1], R.D.GRIGORIEFF [1,2], E.HAIRER, S.P.NØRSETT & G.WANNER [1], P.HENRICI [1], J.D. LAMBERT [2],....), and we shall not bother about existence or uniqueness questions concerning the solutions of the problems under consideration; to speak with H. Stetter, "uniqueness will often be achieved by suitably restricting the domain of the equation" (H.J.STETTER [3]).

3.6.1 One-Step Methods for First-Order Problems. Let us consider the initial value problem for a first order ordinary differential equation

$$y' = f(t,y)\,, \quad y(t_0) = y_0\,, \tag{3.150}$$

where f is assumed to be sufficiently smooth. Suppose we want to compute a numerical approximation of the value $y(x)$, where x is some real number. To do this, we use a one-step method, i.e. a numerical procedure of the form

$$y_{\nu+1} \;:=\; y_\nu + h\,\Phi(t_\nu, y_\nu, h) \tag{3.151}$$

for $\nu = 0, \ldots, n-1$, with

$$h \;:=\; \frac{x - t_0}{n} \quad \text{and} \quad t_\nu = t_0 + \nu\,h\,, \tag{3.152}$$

and take y_n as an approximation for $y(x)$. Everywhere in this section, y_n (i.e., the index n) will denote the final result of a process like (3.151), while y_ν is intermediate value of this process; in other words, $n \cdot h$ is constant as n goes to infinity.

The user-chosen number n is usually denoted as *stepsize parameter*, whereas we shall speak of h as the *stepsize*; the function Φ is the *increment function* of the one-step method (3.151).

It is quite obvious from (3.152) that a sequence, say $\{y_n\}$, possesses an asymptotic expansion in powers of h for $h \to 0$, if and only if it possesses one in (the same) powers of $1/n$ for $n \to \infty$. Therefore we will leave at several places in this and the next section our terminology of the preceeding sections and work with asymptotic expansions in powers of the stepsize h – as almost all people in ODE-theory do.

In the following we shall almost always assume that the stepsize h is *fixed*; however, it should be remarked that most of the results carry over to methods with a variable stepsize, i.e., with (3.151) replaced by

$$y_{\nu+1} := y_\nu + h_\nu \, \Phi(t_\nu, y_\nu, h_\nu) \tag{3.153}$$

with

$$h_\nu > 0 \text{ for all } \nu, \quad \sum_{\nu=0}^{n-1} h_\nu = x - t_0 \quad \text{and} \quad t_\nu := t_0 + \sum_{j=0}^{\nu-1} h_j ,$$

as long as the stepsizes $h_\nu = h_\nu^{(n)}$ behave reasonably well. For example, we might assume the existence of two constants C_1, C_2 , such that for all ν

$$\frac{C_1}{n} \le h_\nu^{(n)} \le \frac{C_2}{n} , \tag{3.154}$$

as n goes to infinity. In this case, the sequence $\{h_\nu^{(n)}\}_{n\in\mathbb{N}}$ is said to be *linearly convergent.*

We shall assume throughout that the increment function Φ and therefore also the numerical procedure (3.151) is sufficiently smooth and *consistent* with the problem (3.150), which can be expressed through the condition

$$\Phi(t, y, 0) = f(t, y) \text{ for all } (t, y) . \tag{3.155}$$

In order to get some information on the *global error* $(y_n - y(x))$ of a method, it is reasonable to study first the so-called *local error*, i.e. that portion of the error that is brought in by going from ν to $(\nu + 1)$ in (3.151). The precise definition of the term local error is by no means unique in the literature (cf. E.HAIRER, S.P.NØRSETT & G.WANNER [1], p. 160, and also – in the case of multistep methods – J.D.LAMBERT [1]); we shall follow the approach of E.HAIRER, S.P.NØRSETT & G.WANNER [1] and define the term

$$l(t, h) := y(t + h) - y(t) - h\,\Phi(t, y(t), h) \tag{3.156}$$

as the *local error (function)* of the method (3.151). Using Taylor expansion, it can easily be shown that the local error of a consistent method is of order $O(h^2)$ at least, that is

$$l(t,h) \;=\; O(h^2) \qquad \text{for } h \to 0 . \tag{3.157}$$

Example. Consider the increment function $\Phi(t,y,h) := f(t,y)$, which defines Euler's method

$$y_{\nu+1} \;:=\; y_\nu + h\, f(t_\nu, y_\nu) , \tag{3.158}$$

and apply it to the first order problem

$$y \;=\; y' , \quad y(0) = 1$$

with the solution $y(t) = \mathrm{e}^t$. The local error function is given by

$$l(t,h) \;=\; \mathrm{e}^t\,(\mathrm{e}^h - (1+h)) ,$$

and we see, using Taylor expansion and (3.152), that $l(t,h)$ possesses the logarithmic asymptotic expansion

$$l(t,h) \;=\; \sum_{\mu=2} \frac{c_\mu(t)}{n^\mu} .$$

This behavior of the local error is typical for almost all one-step methods, in particular for those of Runge-Kutta type (see below); the hope is now that the existence of an asymptotic expansion carries over from the local to the global error. In the present example, this is indeed true and can be proved directly, since after n steps with stepsize $h = x/n$ the approximating value y_n can be written in closed form by $y_n = (1 + x/n)^n$. This is a very interesting function for our purposes and will serve as an example at several places. Therefore we summarize its properties in the following lemma, at the same time closing this example.

Lemma 3.45: *The sequence of functions $\{\sigma_n^e(z)\}$, defined for each $z \in \mathbb{C}$ through*

$$\sigma_n^e(z) := \left(1 + \frac{z}{n}\right)^n, \tag{3.159}$$

possesses a logarithmic asymptotic expansion of arbitrary order with modulus 1 *and limit* e^z, *i.e.*

$$\sigma_n^e(z) = \mathrm{e}^z + \sum_{\mu=1} \frac{c_\mu(z)}{n^\mu} \quad \textit{for } n \to \infty . \tag{3.160}$$

Furthermore, the coefficient functions $c_\mu(z)$ *are of the form*

$$c_\mu(z) = \mathrm{e}^z \cdot \sum_{j=\mu+1}^{2\mu} \alpha_{j,\mu}\, z^j . \tag{3.161}$$

Proof. Applying the natural logarithm and its functional equation to $\sigma_n^e(z)$, we obtain for all $n > |z|$:

$$\begin{aligned} \log(\sigma_n^e(z)) &= n \cdot \log(1 + \frac{z}{n}) \\ &= n \cdot \sum_{\nu=1}^{\infty} (-1)^{\nu+1} \frac{z^\nu}{n^\nu \cdot \nu} = z + S_n(z) \end{aligned}$$

with

$$S_n(z) := \sum_{\nu=2}^{\infty} (-1)^{\nu+1} \frac{z^\nu}{n^{\nu-1} \cdot \nu} . \tag{3.162}$$

To get back to the functions $\sigma_n^e(z)$ themselves, we have to apply the exponential function, which gives us the representation

$$\sigma_n^e(z) = \mathrm{e}^{z+S_n(z)} = \mathrm{e}^z \cdot \left(1 + \sum_{\lambda=1}^{\infty} \frac{S_n(z)^\lambda}{\lambda!}\right),$$

and we can read off the assertions of Lemma 3.45. □

It was already pointed out in Subsection 3.3.2 that one of the most important matrix functions is surely the *matrix exponential*; therefore we remark here that the logarithmic asymptotic expansion (3.160) also holds, if the scalar variable z is replaced by some arbitrary square matrix A, see G.WALZ [2].

Of course it cannot be expected in general that there is a closed-form representation of y_n, and therefore a direct proof like the one above (which does not make use of the local error at all) is very exceptional. However, there is the following fundamental Theorem 3.46, which says that the existence of an asymptotic expansion of the local error always implies the existence of an asymptotic expansion of the global error. The first proof of this result, which was quite long and difficult (but remember that the first proof of an interesting result may be as ugly as it wants to be!), is due to W.B.GRAGG [1]; it was worked out further and simplified in H.J.STETTER [3], and the form of the statement as well as the proof below is due to E.HAIRER & CH.LUBICH [1]. In my opinion, it is the most beautiful and at the same time straightforward approach that is possible; furthermore, it has the advantage of being generalizable to the multistep case, see Subsection 3.6.3.

Theorem 3.46 (W.B.GRAGG [1], E.HAIRER & CH.LUBICH [1]): *In addition to the assumptions from above, suppose that the local error function of the one-step method (3.151) possesses an asymptotic expansion of order* $m+1$ *of the form*

$$l(t,h) \;=\; \sum_{\mu=p+1}^{m+1} d_\mu(t)\cdot h^\mu \;+ O(h^{m+2}) \quad \text{for } h\to 0\,, \qquad (3.163)$$

where m *is some integer greater than or equal to* p.
Then the global error possesses the following asymptotic expansion of order m

$$y_n \;=\; y(x) \;+\; \sum_{\mu=p}^{m} e_\mu(x)\cdot h^\mu \;+ o(h^m) \quad \text{for } h\to 0\,. \qquad (3.164)$$

(Note that we have $p\geq 1$ *in all cases, due to (3.157)).*

Remark. Since every "well-behaved" one-step method of order p possesses a local expansion of the type (3.163), we have here a very powerful theorem with a wide range of applications.

Proof of Theorem 3.46. The proof is by induction with respect to μ, so let us start with $\mu = p$ and define a function e_p as the (unique) solution of the initial value problem

$$e_p'(t) = \frac{\partial}{\partial y} f(t, y(t))\, e_p(t) - d_{p+1}(t)\,, \quad e_p(t_0) = 0\,. \tag{3.165}$$

With this function e_p, we consider now a new increment function, say $\Phi^{(1)}$, defined by

$$\Phi^{(1)}(t, y(t), h) := \Phi\big(t, y(t) + e_p(t)h^p, h\big) - \big(e_p(t+h) - e_p(t)\big)h^{p-1}\,, \tag{3.166}$$

and the hereby also defined one-step method

$$y_{\nu+1}^{(1)} := y_\nu^{(1)} + h\,\Phi^{(1)}(t_\nu, y_\nu^{(1)}, h)\,,\; y_0^{(1)} := y_0\,. \tag{3.167}$$

Expanding the local error of this method, say $l^{(1)}(t,h)$, into powers of h we obtain

$$\begin{aligned} l^{(1)}(t,h) &= y(t+h) - y(t) - h\,\Phi^{(1)}(t, y(t), h) \qquad (3.168)\\ &= \left[d_{p+1}(t) - \frac{\partial}{\partial y}\Phi(t,y,0)e_p(t) + e_p'(t)\right] h^{p+1} + O(h^{p+2}), \end{aligned}$$

where we have used (3.163) and

$$\frac{\partial}{\partial y}\Phi(t,y,h) = \frac{\partial}{\partial y}\Phi(t,y,0) + O(h)$$

(remember that Φ was assumed to be sufficiently differentiable).

Since e_p is the solution of the initial value problem (3.165), we see, using

$$\frac{\partial}{\partial y}\Phi(t,y,0) = \frac{\partial}{\partial y} f(t,y)$$

that the h^{p+1}-term in (3.168) vanishes. Therefore, the local error $l^{(1)}(t,h)$ is of order $O(h^{p+2})$, and a standard result from ODE-theory implies that the method defined by $\Phi^{(1)}$ converges and is of order $O(h^{p+1})$, i.e.

$$y_n^{(1)} - y(x) = O(h^{p+1}) = o(h^p). \tag{3.169}$$

Comparing now equation (3.151) with (3.167) we see that (3.166) implies the relation

$$y_n^{(1)} = y_n - e_p(x)h^p .$$

Therefore, from (3.169),

$$y_n = y(x) + e_p(x)h^p + o(h^p),$$

and the first term of the proposed asymptotic expansion has been determined.

If $m = p$, the proof is complete. Otherwise, we repeat the procedure from above, thus defining new increment functions $\Phi^{(2)}, \Phi^{(3)}, \ldots$, until the full asymptotic expansion in (3.164) is established. □

A very attractive situation is given if the asymptotic expansion in (3.164) has modulus 2, i.e. if it contains only even powers of h. This case, in which we speak of an h^2– *expansion,* is a very desirable one, since here the application of extrapolation methods yields very efficient algorithms. If we ask for criteria, under which a one-step method possesses an h^2– expansion, the magic key word *symmetry* appears everywhere in the literature, i.e., symmetric methods possess h^2– expansions.

However, although this fact was known intuitively for quite a long time, its precise proof, and in particular an exact definition of the term symmetry in this context turned out to be a hard piece of work. The approach we are going to present is based mainly on the work of E.HAIRER, S.P.NØRSETT & G.WANNER [1] and H.J.STETTER [2,3].

Let us rewrite the one-step method (3.151) in the form

$$y(t+h) = y(t) + h\,\Phi(t, y(t), h)\,. \tag{3.170}$$

Definition 3.47: We call the one-step method (3.170) *symmetric,* if the increment function Φ keeps invariant after replacing h by $-h$ and then t by $t+h$, i.e. if we have formally

$$\Phi(t, y(t), h) = \Phi(t+h, y(t+h), -h)\,. \tag{3.171}$$

There exist several motivations for this definition of the term symmetry. A first and more intuitive one is the following: If we consider a method whose increment function satisfies (3.171), then "it does not matter" if we step forward (from t to $t+h$) with stepsize h, or backward (from $t+h$ to t) with stepsize $-h$, we always end up with the same method.

Another, more formal approach to our definition of symmetry is the following one: Starting with (3.170), let us replace h by $-h$ and then t by $t+h$; we obtain the procedure

$$y(t) = y(t+h) - h\,\Phi(t+h, y(t+h), -h)\,. \tag{3.172}$$

Of course, (3.172) constitutes a one-step (backward) method of its own (in the terminology of E.HAIRER, S.P.NØRSETT & G.WANNER [1] it is the *adjoint* method of (3.170)), and if (3.171) holds, we see that both methods conincide.

A slightly different definition of symmetry, sometimes more suitable for testing a method to be symmetric, is the following one, due to H.J.STETTER [2]: Suppose that our one-step method (3.151) can be written in the form

$$y_{\nu+1} = y_\nu + h\,\Psi(t_\nu, t_{\nu+1}, y_\nu, y_{\nu+1}, h) \tag{3.173a}$$

where the increment function Ψ satisfies

$$\Psi(\xi_1, \xi_2, \eta_1, \eta_2, h) \;=\; \Psi(\xi_2, \xi_1, \eta_2, \eta_1, -h) \qquad (3.173b)$$

for all $\xi_1, \xi_2, \eta_1, \eta_2, h$ under consideration. Then this method is called symmetric. Obviously, for $t = t_\nu$ this definition of symmetry implies the one above.

As a first example, we consider the *trapezoidal rule*

$$y_{\nu+1} \;=\; y_\nu + h \cdot \frac{f(t_\nu, y_\nu) + f(t_{\nu+1}, y_{\nu+1})}{2} \qquad (3.174)$$

and the *implicit midpoint rule*

$$y_{\nu+1} \;=\; y_\nu + h\, f\Big(\frac{t_\nu + t_{\nu+1}}{2}, \frac{y_\nu + y_{\nu+1}}{2}\Big), \qquad (3.175)$$

which are both easily seen to satisfy (3.173) and therefore to be symmetric.

The following Theorem 3.48 indicates why symmetry is such a desirable property in the context of asymptotic expansions and extrapolation methods. Once more, this result appears in several forms in a number of publications, such as W.B.GRAGG [1,2], E.HAIRER & CH.LUBICH [1], E.HAIRER, S.P.NØRSETT & G.WANNER [1] or H.J.STETTER [2,3]:

Theorem 3.48: *In addition to the assumptions of Theorem 3.46, let the one-step method under consideration be symmetric. Then the asymptotic expansion (3.164) of the global error is an h^2-- expansion, i.e. it is of the form*

$$y_n(x) \;=\; y(x) \;+\; \sum_{j=0}^{\tilde{m}} e_{p+2j}(x) \cdot h^{p+2j} \;+ o(h^{p+2\tilde{m}}) \;\; \text{for } h \to 0 \qquad (3.176)$$

with p even and

$$p + 2\tilde{m} \;=\; \begin{cases} m, & \text{if } m \text{ even}, \\ m-1, & \text{if } m \text{ odd}. \end{cases}$$

Proof. Re-examining the proof of Theorem 3.46 we recognize that the final result of the method

$$y(t+h) \;=\; y(t) + h\,\Phi(t+h, y(t+h), -h)\,,$$

say y_{-n}, possesses the same asymptotic expansion as the original one, given in (3.170), but with h replaced by $-h$. Since in the symmetric case both methods coincide, the theorem follows. □

To give a first illustration of Theorem 3.48, let us continue our example from the beginning of this section, concerning the initial value problem

$$y \;=\; y'\,, \quad y(0) \;=\; 1\,.$$

This time, we apply the trapezoidal rule (3.174), which gives us, after n steps with stepsize $h = x/n$, the final result

$$y_n \;=\; y_n(x) \;=\; \left(\frac{1+\frac{x}{2n}}{1-\frac{x}{2n}}\right)^n \tag{3.177}$$

in closed form.

Using again the functions $\sigma_n^e(x) = (1+\frac{x}{n})^n$ (see (3.159)), we may also write

$$y_n(x) \;=\; \sigma_n^e(x/2)\cdot\sigma_{-n}^e(x/2)\,, \tag{3.178}$$

and from Lemma 3.45 it follows that $y_n(x)$ possesses the logarithmic asymptotic expansion

$$\begin{aligned} y_n(x) \;&=\; \left(e^{\frac{x}{2}} + \sum_{\mu=1} \frac{c_\mu(\frac{x}{2})}{n^\mu}\right)\cdot\left(e^{\frac{x}{2}} + \sum_{\mu=1} \frac{(-1)^\mu c_\mu(\frac{x}{2})}{n^\mu}\right) \\ &=:\; e^x \;+\; \sum_{\mu=1} \frac{d_\mu(x)}{n^\mu} \end{aligned} \tag{3.179}$$

with certain coefficient functions d_μ. Moreover, since $y_n(x) = y_{-n}(x)$ for all n and x, in (3.179) all d_μ with odd index must vanish, and we have finally proved that

$$y_n(x) = e^x + \sum_{j=0} \frac{d_{2+2j}(x)}{n^{2+2j}} ,$$

in accordance with Theorem 3.48. □

In the next subsection, we shall study two important classes of one-step methods, to which the above results apply.

3.6.2 Runge-Kutta Methods and Generalized Trapezoidal Rules. There are two obvious ways to improve Euler's method (3.158), the simplest numerical procedure for solving an initial value problem for a first order differential equation. One is, to take into account – for the computation of the new value $y_{\nu+1}$ – not only the information that comes from y_ν, but also that from $y_{\nu-1}, y_{\nu-2}$, and so on. This leads to (in most applications linear) multistep methods, which we will study in Subsection 3.6.3.

The other possibility is to stay with the one-step form, but to go to more complicated (and therefore more promising) increment functions Φ. This will be our subject in the present subsection.

First we follow an idea of C.RUNGE [1,2], K.HEUN [1] and W.KUTTA [1] and define, for some natural number s, the following type of one-step method for the solution of the initial value problem (3.150), nowadays well-known as *Runge-Kutta method with s stages*: With some real coefficients a_{ij}, b_i, c_i, to be determined later, the νth step of the method is defined as

$$y_{\nu+1} = y_\nu + h \sum_{i=1}^{s} b_i k_i , \tag{3.180a}$$

where, for $i = 1, \ldots, s$,

$$k_i = k_i(t_\nu, y_\nu, h) = f\Big(t_\nu + c_i h,\, y_\nu + h \sum_{j=1}^{s} a_{ij} k_j\Big) . \tag{3.180b}$$

It is convenient to assume that

$$\sum_{j=1}^{s} a_{ij} = c_i \tag{3.181}$$

for all i. Furthermore, we can easily verify that the Runge-Kutta method (3.180) is consistent, if and only if

$$\sum_{i=1}^{s} b_i = 1 \tag{3.182}$$

holds. From now on we shall always assume conditions (3.181) and (3.182) to be satisfied.

We have to bring into the game some more notations and definitions: The Runge-Kutta method is called *explicit*, if

$$a_{ij} = 0 \quad \text{for } j \geq i\,,\ i = 1, \ldots, s\,. \tag{3.183}$$

Until the 1960's, almost exclusively Runge-Kutta methods of this type were used and investigated. The reason is obvious: If Condition (3.183) is satisfied, then each k_i in (3.180) can be computed explicitly by means of the previously calculated values $k_1, \ldots, k_{i-1}$. In all other cases, i.e. if there happens some a_{ij} with $j \geq i$ to be different from zero (in which case we speak of an *implicit* Runge-Kutta method), it is necessary in each step to solve an in general nonlinear implicit system of equations for the computation of k_i. This implies high computational costs and therefore implicit Runge-Kutta methods cannot compete with explicit ones, at least as long as we are dealing with non-stiff equations.

Finally we remark that some authors also speak of *semi-implicit* (or *semi-explicit*, depending on the question if the author has a more pessimistic or optimistic view of things) Runge-Kutta methods, if condition (3.183) is weakened to be

$$a_{ij} = 0 \quad \text{for } j > i\,,\ i = 1, \ldots, s\,.$$

There is a very nice and instructive interpretation of Runge-Kutta methods, due to J.D. Lambert. The following lines are quoted (with some obvious changes of notation) from J.D.LAMBERT [2], p. 151:

Runge-Kutta methods (...) constitute a clever and sensible idea. The unique solution of a well-posed initial value problem can be thought of as a single integral curve in $\mathbb{R}^2$, but, due to truncation and round-off error, any numerical solution is, in effect, going to wander off that integral curve, and the numerical solution is inevitably going to be affected by the behaviour of neighbouring integral curves. Thus it is the behaviour of the *family* of integral curves, and not just that of the unique solution curve, that is of importance (...). Runge-Kutta methods deliberately try to gather information about this family of curves. Such an interpretation is most easily seen in the case of explicit Runge-Kutta methods, where, from (3.180), (3.181) and (3.183), we have

$$\begin{aligned} k_1 &= f(t_\nu, y_\nu), \\ k_2 &= f(t_\nu + c_2 h,\ y_\nu + h c_2 k_1), \\ k_3 &= f(t_\nu + c_3 h,\ y_\nu + h(c_3 - a_{32})k_1 + h a_{32} k_2), \\ &\ \vdots \end{aligned}$$

Start from (t_ν, y_ν) , take one step of Euler's method of length hc_2 and evaluate the derivative vector at the point so reached; the result is k_2 . We now have two samples for the derivative, k_1 and k_2 , so let us take the weighted mean of k_1 and k_2 as the initial slope in another Euler step of length hc_3 , and evaluate the derivative at the point so reached; the result is k_3 . Continuing in this manner, we obtain a set k_i , $i = 1, \ldots, s$ of samples of the derivative. The final step is yet another Euler step

$$y_{\nu+1} = y_\nu + h \sum_{i=1}^{s} b_i k_i$$

from (t_ν, y_ν) to $(t_{\nu+1}, y_{\nu+1})$, using as initial slope the weighted mean of the samples $k_1, \ldots, k_s$. Thus, an explicit Runge-Kutta method sends out feelers into the solution space, to gather samples of the derivative, before deciding in which direction to take an Euler step, an eminently sensible idea.

It is convenient to define the order of a Runge-Kutta method by comparing the Taylor expansion of the true solution with that of the increment function of the method. If we assume the true solution y of the problem to be sufficiently smooth, we can write down its Taylor expansion

$$y(t+h) \;=\; y(t) + h\,T(t,y,h) \;+\; O(h^{K+1}) \;\text{ for }\; h \to 0\,, \tag{3.184a}$$

where

$$T(t,y,h) \;:=\; \sum_{\kappa=1}^{K} \frac{y^{(\kappa)}(t)}{\kappa!}\,h^{\kappa-1} \;=\; \sum_{\kappa=1}^{K} \frac{f^{(\kappa-1)}(t,y)}{\kappa!}\,h^{\kappa-1}\;. \tag{3.184b}$$

Here, K is a natural number depending on the smoothness of f, which we assume to be sufficiently large, and $f^{(\kappa-1)}$ stands for $\frac{\partial^{(\kappa-1)}}{\partial t^{(\kappa-1)}} f$.

Definition 3.49: The Runge-Kutta method (3.180) is of *order* p, if the Taylor expansion of the increment function $\sum_{i=1}^{s} b_i k_i$ agrees with T from (3.184) up to the h^{p-1}– term, i.e. if

$$\sum_{i=1}^{s} b_i k_i \;-\; T(t,y,h) \;=\; \sum_{\lambda=p}^{\tilde{K}} \gamma_\lambda\, h^\lambda \;+\; O(h^{\tilde{K}+1}) \;\text{ for }\; h \to 0 \tag{3.185}$$

with some $\tilde{K} \le K$, again depending (through the $k_i's$) on the smoothness of f.

In a series of papers, the main results of which are summarized in J.C.BUTCHER [1], this author investigated (and to a large extend answered) the question of the maximal order p_{max} that can be attained by an $s-$ stage explicit Runge-Kutta method. We cannot reproduce his long and difficult proofs here, but state the result, the so-called *Butcher barriers*, in the following theorem.

Theorem 3.50 (J.C.BUTCHER [1,3]): *Let*

$$p_{max}(s) \;:=\; \max\{p \in \mathbb{N} \mid \text{there exists an explicit } s\text{–stage Runge-Kutta method of order } p\}\;.$$

Then we have

$$p_{max}(s) \;=\; \begin{cases} s\,, & \textit{for } s = 1,2,3,4, \\ s-1\,, & \textit{for } s = 5,6,7, \\ s-2\,, & \textit{for } s = 8,9, \\ s-3\,, & \textit{for } s = 10. \end{cases}$$

Futhermore, for $s \geq 11$, *the estimation*

$$p_{max}(s) \;\leq\; s-3$$

holds.

It is already obvious from Definition 3.49 that a $p^{th}-$order Runge-Kutta method leads to a numerical approximation of the true solution with an asymptotic expansion in powers of h , beginning with an h^p- term; this is worked out in the next theorem.

Theorem 3.51: *In the notation of the previous subsection, let* $y_n = y_n(x)$ *denote the result of a Runge-Kutta method of order* p , *applied to the initial value problem (3.150) with sufficiently smooth function* f . *Then the global error of this method possesses an asymptotic expansion of the form*

$$y_n(x) \;=\; y(x) \;+\; \sum_{\mu=p}^{\tilde{K}} e_\mu(x)\cdot h^\mu \;+ o(h^{\tilde{K}}) \qquad \textit{for } h\to 0\,. \tag{3.186}$$

Proof. We want to apply Theorem 3.46, and therefore we have to investigate the local error function

$$l(t,h) \;=\; y(t+h) - y(t) - h\sum_{i=1}^{s} b_i k_i$$

of our method. Using the relations (3.185) and (3.184a), we obtain

$$\begin{aligned} l(t,h) \;&=\; y(t+h) - y(t) - h\Big(T(t,y,h) + \sum_{\lambda=p}^{\tilde{K}} \gamma_\lambda\, h^\lambda + O(h^{\tilde{K}+1})\Big) \\ &=\; \sum_{\lambda=p+1}^{\tilde{K}+1} \gamma_\lambda\, h^\lambda + O(h^{\tilde{K}+2})\,, \end{aligned}$$

and the result follows from Theorem 3.46. □

Example. For $s = 1$ the good old Euler method is the only explicit Runge-Kutta method that is possible, so let us turn immediately to the case $s = 2$. We have, using (3.181) and (3.182),

$$\begin{aligned} T(t,y,h) &= f(t,y) + \frac{h}{2}\left(\frac{\partial}{\partial t} f(t,y) + f(t,y)\frac{\partial}{\partial y} f(t,y)\right) + \\ &\qquad + \left\{\text{higher order terms}\right\} \end{aligned}$$

and

$$\begin{aligned} b_1k_1 + b_2k_2 &= (1-b_2)f(t,y) + b_2 f(t,y) + \\ &\qquad + b_2c_2h\left(\frac{\partial}{\partial t} f(t,y) + f(t,y)\frac{\partial}{\partial y} f(t,y)\right) + \\ &\qquad + \left\{\text{higher order terms}\right\}. \end{aligned}$$

It follows that for each real number $\alpha \neq 0$ the 2-stage Runge-Kutta method with increment function

$$(1-\alpha)f(t,y) + \alpha\, f(t + \frac{h}{2\alpha}, y + \frac{h}{2\alpha} f(t,y)) \tag{3.187}$$

is of order 2 and possesses an asymptotic expansion of the form (3.186) with $p = 2$.

Let us illustrate this once again with our prototype problem

$$y' = y, \quad y(0) = 1.$$

In this case, the increment function (3.187) takes for each α just the form $y(1 + h/2)$, thus our Runge-Kutta method reads

$$y_{\nu+1} = y_\nu(1 + h + h^2/2),$$

and leads to the result (remember that $h = x/n$)

$$y_n(x) \;=\; \left(1 + \frac{x}{n} + \frac{x^2}{2n^2}\right)^n . \qquad (3.188)$$

With exactly the same proof as in Lemma 3.45, it follows that the sequence $\{y_n(x)\}$ possesses the asymptotic expansion

$$y_n(x) \;=\; \mathrm{e}^x \;+\; \sum_{\mu=2} \frac{c_\mu(x)}{n^\mu} .$$

One might ask now if there exist Runge-Kutta methods with expansions in even powers of h. In the case of explicit ones the answer is most probably negative, as follows from the next result.

Theorem 3.52 (G.Wanner [1], H.J.Stetter [3]): *The Runge-Kutta method (3.180) is symmetric, if for* $i,j = 1,\ldots,s$ *the relations*

$$a_{s+1-i,s+1-j} + a_{ij} \;=\; b_{s+1-j} \;=\; b_j \qquad (3.189)$$

hold.

Due to (3.183), Theorem 3.52 has the following immediate implication.

Corollary 3.53: *There is no symmetric explicit Runge-Kutta method.*

Although symmetry was only proved to be a sufficient condition for a method to have a h^2- expansion, these two properties are so close together that in my opinion there is no hope to find an explicit Runge-Kutta method with an h^2- expansion.

We now turn to another class of interesting one-step methods, which has been introduced and investigated quite recently; all methods in this class enjoy the property of being symmetric and therefore lead to h^2- expansions.

Let us start with the classical trapezoidal rule (cf. (3.174)), which we write now in the form

$$y_{\nu+1} \;=\; y_\nu + h\,\Psi^T(t_\nu, t_{\nu+1}, y_\nu, y_{\nu+1}, h) \qquad (3.190a)$$

with the increment function

$$\Psi^T := \Psi^T(t_\nu, t_{\nu+1}, y_\nu, y_{\nu+1}, h) := \frac{f(t_\nu, y_\nu) + f(t_{\nu+1}, y_{\nu+1})}{2} \tag{3.190b}$$

This method was already recognized to be symmetric in the sense of (3.173).

The idea is now to replace the Arithmetic mean on the right-hand side of (3.190b) by other – in general nonlinear – means, which gave the resulting one-step methods the name *generalized trapezoidal rules.* We consider the following types of means resp. increment functions:

$$\Psi^G := \sqrt{f_\nu \, f_{\nu+1}} \quad \text{(Geometric)}\,, \tag{3.191a}$$

$$\Psi^H := \frac{2\, f_\nu\, f_{\nu+1}}{f_\nu + f_{\nu+1}} \quad \text{(Harmonic)}\,, \tag{3.191b}$$

$$\Psi^L := \frac{f_{\nu+1} - f_\nu}{\log(f_{\nu+1}/f_\nu)} \quad \text{(Logarithmic)}\,, \tag{3.191c}$$

$$\Psi^* := \frac{f_\nu^2 + f_{\nu+1}^2}{f_\nu + f_{\nu+1}} \quad \text{(Contra-Harmonic)}\,. \tag{3.191d}$$

Here we have used the abbreviations $f_\nu := f(t_\nu, y_\nu)$ and $f_{\nu+1} := f(t_{\nu+1}, y_{\nu+1})$. Of course, in certain cases some additional assumptions on the f_ν 's (e.g. positivity) have to be imposed, which we will always do in the following. The behavior of these one-step methods has been studied extensively in a series of papers by D.J.EVANS & B.B.SANUGI [1,2,3] and D.J.EVANS & G.WALZ [1]. The main result concerning asymptotic expansions of these investigations is summarized in the next theorem; it turns out that each of these methods is of order 2 and possesses an h^2– expansion.

Theorem 3.54 (D.J.Evans & G.Walz [1]): *Let Ψ denote an arbitrary one of the increment functions defined in (3.191), and consider the hereby defined one-step method*

$$y_{\nu+1} = y_\nu + h\,\Psi(t_\nu, t_{\nu+1}, y_\nu, y_{\nu+1}, h)\,, \quad \nu = 0, \ldots, n-1, \quad (3.192)$$

with $h = (x - t_0)/n$ and $t_\nu = t_0 + \nu h$ for $\nu = 1, \ldots, n$.

Then the global error of this method possesses an asymptotic expansion of the form

$$y_n(x) = y(x) + \sum_{\mu=1}^{M} c_{2\mu}(x) \cdot h^{2\mu} + o(h^{2M}) \quad \text{for } h \to 0\,. \quad (3.193)$$

The order M of this expansion depends on the smoothness properties of f resp. Ψ.

Proof. We apply Theorems 3.46 and 3.48. The consistency condition (3.155) as well as the symmetry condition (3.173) are immediately seen to be satisfied for each of the four increment functions in (3.191), due to the properties of a mean.
So, we are left to prove the existence of an expansion like (3.163) for the local error functions. We begin with the geometric mean method and find, using Taylor expansion at t that

$$\begin{aligned} \Psi^G(t, t+h, y(t), y(t+h), h) &= \\ &= \sqrt{y'(t)\, y'(t+h)} \\ &= \sqrt{y'(t) \cdot \sum_{j=0}^{K_0} \frac{h^j}{j!}\, y^{(j+1)}(t) + O(h^{K_0+1})} \\ &= y'(t) \cdot \sqrt{1 + \sum_{j=1}^{K_0} \frac{h^j}{j!}\, \frac{y^{(j+1)}(t)}{y'(t)} + O(h^{K_0+1})} \qquad (3.194) \end{aligned}$$

with some $K_0 \in \mathbb{N}$. If h is small enough, the radicand in (3.194) is of the form $(1+\xi)$ with $|\xi| < 1$, such that we can use the series expansion of the square root function and find that (3.194) is of the form

$$y'(t) + \frac{h}{2}\, y''(t) + \sum_{\mu=3}^{K_1} a_\mu(t) h^{\mu-1} + O(h^{K_1}) \tag{3.195}$$

with certain coefficients a_μ and some $K_1 \in \mathbb{N}$.

On the other hand, we have

$$y(t+h) - y(t) \;=\; h\, y'(t) + \frac{h^2}{2}\, y''(t) + \sum_{\mu=3}^{K_2} b_\mu(t) h^\mu + O(h^{K_2+1})\,, \tag{3.196}$$

and a combination of (3.195) and (3.196) shows us that the local error function of the geometric mean formula,

$$l^G(t,h) \;=\; y(t+h) - y(t) - h\, \Psi^G(t, t+h, y(t), y(t+h), h)\,,$$

possesses the expansion

$$l^G(t,h) \;=\; \sum_{\mu=3}^{M+1} d_\mu(t) \cdot h^\mu \; + O(h^{M+2}) \quad \text{for } h \to 0$$

with some $M \in \mathbb{N}$, depending on K_0, K_1, K_2, i.e. on the smoothness of f. Finally, an application of Theorems 3.46 and 3.48 proves Theorem 3.54 for the case of the Geometric mean method.

As for the Harmonic mean method, the proof runs along similar lines as the one just examined, and so we can be brief: Using the abbreviations $y^{(\mu)} := y^{(\mu)}(t)$ for all μ, we may again apply Taylor's formula to the increment function Ψ^H, which yields

$$\Psi^H(t, t+h, y(t), y(t+h), h) \;=$$

$$= \; y' \cdot \frac{1 + h\,\frac{y''}{y'} + \frac{h^2}{2}\frac{y'''}{y'} + \sum_{\mu=3}^{J_0} \alpha_\mu h^\mu + O(h^{J_0+1})}{1 + \frac{h}{2}\frac{y''}{y'} + \frac{h^2}{4}\frac{y'''}{y'} + \sum_{\mu=3}^{J_0} \tilde{\alpha}_\mu h^\mu + O(h^{J_0+1})}\,. \tag{3.197}$$

If h is small enough, we can expand the right hand side of (3.197) into a geometric series, thus obtaining that Ψ^H is of the form

$$y'\left(1+h\,\frac{y''}{y'}+\frac{h^2}{2}\frac{y'''}{y'}+\sum_{\mu=3}^{J_0}\alpha_\mu h^\mu+O(h^{J_0+1})\right)\cdot$$

$$\cdot\left(1-\frac{h}{2}\frac{y''}{y'}-\frac{h^2}{4}\frac{y'''}{y'}+\frac{h^2}{4}\frac{(y'')^2}{(y')^2}+\sum_{\mu=3}^{J_1}\beta_\mu h^\mu+O(h^{J_1+1})\right)$$

$$= y'+\frac{h}{2}y''+\frac{h^2}{4}\left(y'''-\frac{(y'')^2}{y'}\right)+\sum_{\mu=3}^{J_2}\delta_\mu h^\mu+O(h^{J_2+1})\,.$$

Finally, we use again the representation (3.196) and recognize that the local error of the Harmonic mean method possesses an asymptotic expansion of the desired type.

The proof of the corresponding results for the Logarithmic and the Contra-Harmonic mean should now be clear (just use the series expansion of the logarithm and the trick of expanding the denominators into a geometric series) and is therefore omitted here. □

At the end of this section, we would like to show now – by means of two specific examples – that once again one cannot expect to obtain h^2– expansions in general, as soon as non-symmetric formulae are used.

Our first example is concerned with another generalization of the classical trapezoidal formula, sometimes also denoted as ϑ– method, defined through the increment function

$$\Phi^\vartheta(t_\nu,t_{\nu+1},y_\nu,y_{\nu+1},h) := \vartheta f_\nu+(1-\vartheta)\,f_{\nu+1}\,,\quad 0\le\vartheta\le 1 \text{ fixed.}$$

Application of this method to the prototype initial value problem

$$y'(t)=y(t)\,,\quad y(0)=1$$

with the exact solution $y(t) = \exp(t)$ leads to the one-step method

$$y_{\nu+1} = y_\nu + h(\vartheta f_\nu + (1-\vartheta) f_{\nu+1}), \quad \nu = 0, \ldots, n-1.$$

The final result of this method, which we denote by $y_n(\vartheta, x)$, i.e. the numerical approximation of the theoretical value $y(x)$, can be written in closed-form, it is

$$\begin{aligned} y_n(\vartheta, x) &= \left(\frac{1 + (\vartheta x)/n}{1 - ((1-\vartheta)x)/n} \right)^n \\ &= \sigma_n^e(\vartheta x) \cdot \sigma_{-n}^e((1-\vartheta)x) \end{aligned} \tag{3.198}$$

with $\sigma_n^e(z) = (1 + z/n)^n$.

Using Lemma 3.45, we see after some straightforwad calculations that (3.198) implies the existence of the asymptotic expansion

$$y_n(\vartheta, x) = \exp(x) \left(1 + \sum_{\mu=1} \frac{d_\mu(\vartheta, x)}{n^\mu} \right) \tag{3.199}$$

with in general non-vanishing coefficient functions d_μ (if $\vartheta \neq \frac{1}{2}$); for example, it is

$$d_1(\vartheta, x) = \alpha_{2,1} \cdot (2\vartheta - 1) \neq 0 \quad \text{for } \vartheta \neq \frac{1}{2}$$

with $\alpha_{2,1}$ from (3.161). For $\vartheta = \frac{1}{2}$ the classical trapezoidal rule is recovered; in this case it follows from (3.198) that

$$y_n(1/2, x) = y_{-n}(1/2, x) \quad \text{for all } x,$$

i.e. the expansion in (3.199) cannot contain odd powers of n, in accordance with our previous investigations.

In the second example, we consider a certain variable stepsize method, which is a non-symmetric form of the Geometric mean formula from above, and apply it to the differential equation of the logarithm.

For some $x > 1$ and $\nu = 0, \ldots, n$, we define the knots $t_\nu := x^{\frac{\nu}{n}}$, which are linearly convergent in the sense of (3.154). With some arbitrary but fixed α, $0 \le \alpha \le 1$, we consider the (variable stepsize) one-step method

$$y_{\nu+1} := y_\nu + (t_{\nu+1} - t_\nu) \cdot f_\nu^\alpha \cdot f_{\nu+1}^{1-\alpha}, \quad \nu = 0, \ldots, n-1, \qquad (3.200)$$

and apply it to the initial value problem

$$y'(t) = \frac{1}{t}, \quad y(1) = 0$$

with the exact solution $y(t) = \log(t)$. Also in this case, the term y_n can be written in closed form, it is

$$y_n = y_n(\alpha, x) = n\left(x^{\alpha/n} - x^{(\alpha-1)n}\right) \qquad (3.201)$$

and once again the global error possesses an h^2- expansion if and only if the formula is symmetric, i.e. for the geometric mean rule:

Theorem 3.55 (G.Walz [1], D.J.Evans & G.Walz [1]): *For all $x > 1$, the global error of the method (3.200) possesses an asymptotic expansion of the form*

$$y_n(\alpha, x) = y(x) + \sum_{\mu=1} \frac{c_\mu(\alpha, x)}{n^\mu}$$

of arbitrary order.
This expansion contains only even powers of n^{-1}, if and only if $\alpha = 1/2$, i.e. in the symmetric case.

Proof. Due to the relation

$$x^{\alpha/n} = \exp\left(\frac{\alpha \log(x)}{n}\right) = \sum_{j=0}^{\infty} \frac{(\alpha \log(x))^j}{j!} \cdot \frac{1}{n^j},$$

we obtain from (3.201) the representation

$$\begin{aligned} y_n(\alpha, x) &= n \cdot \left(\sum_{j=0}^{\infty} \frac{(\alpha \log(x))^j - ((\alpha-1)\log(x))^j}{j!} \cdot \frac{1}{n^j} \right) \\ &= \log(x) + \sum_{j=2}^{\infty} \frac{\log(x)^j}{j!} \cdot \frac{\alpha^j - (\alpha-1)^j}{n^{j-1}} \\ &=: \log(x) + \sum_{\mu=1} \frac{c_\mu(\alpha, x)}{n^\mu} \end{aligned}$$

$$\text{with } c_\mu(\alpha, x) = \frac{\log(x)^{\mu+1}}{(\mu+1)!} \cdot \left(\alpha^{\mu+1} - (\alpha-1)^{\mu+1}\right),$$

from which the assertion of Theorem 3.55 can be read off. □

3.6.3 Linear Multistep Methods. We had already pointed out in the introduction of Subsection 3.6.2 that there is another natural way to improve Euler's method (3.158), namely to keep the linearity, but to take into account more than one of the previously computed values y_ν. This approach leads to linear multistep methods, which will be considered now.

Let us fix some natural number k. Using from now on the shortened notation

$$f_j := f(t_j, y_j) \quad \text{for } j = 0, \ldots, n, \tag{3.202}$$

a *linear multistep method* or, more precisely, *linear $k-$ step method* for the numerical solution of the initial value problem (3.150) is defined through the difference equation

$$\sum_{j=0}^{k} \alpha_j \, y_{\nu+j} \;=\; h \cdot \sum_{j=0}^{k} \beta_j \, f_{\nu+j} \;, \tag{3.203}$$

which we want to be normalized by the conditions

$$\alpha_k = 1 \quad \text{and} \quad |\alpha_0| + |\beta_0| > 0 \,. \tag{3.204}$$

Here again h denotes the fixed stepsize $h = \frac{x-t_0}{n}$ (cf. (3.152)), and $\{\alpha_j\}_{j=0,\ldots,k-1}$ and $\{\beta_j\}_{j=0,\ldots,k}$ are sets of real parameters.

The choice of suitable parameters is of course a big problem for practical applications, but this cannot be treated here. For example, the α_j 's and β_j 's can be determined via polynomial interpolation, see any textbook on ordinary differential equations, e.g. J.D.LAMBERT [2] or G.MEINARDUS & G.MERZ [2].

It is assumed that the values $y_\nu, \ldots, y_{\nu+k-1}$ are already computed; one single step of the method defined by (3.203) consists now in the computation of the new value $y_{\nu+k}$ by means of this relation. Since $f_{\nu+k}$ depends on $y_{\nu+k}$, the method is called *explicit*, if $\beta_k = 0$, otherwise *implicit.*

Furthermore it is clear that a $k-$ step method is not able to compute the starting values $y_1, \ldots, y_{k-1}$; this must therefore be accomplished by a so-called *starting procedure*, e.g. by one of the one-step-methods discussed in the previous subsection, or by a Taylor-series approach.

Always keeping in mind the aim of our book, we now want to head as directly as possible for Theorems 3.63, 3.64 and 3.66, the main results in this context. However, before coming to that we will have to collect at least some basic facts on such methods.

For the investigation of linear multistep method (3.203) it is convenient to consider the *characteristic* or *generating polynomials*

$$\varrho(z) := \sum_{j=0}^{k} \alpha_j z^j \quad \text{and} \quad \sigma(z) := \sum_{j=0}^{k} \beta_j z^j \tag{3.205}$$

of the method resp. the difference equation (3.203). Here, z denotes a complex dummy variable. It follows from the conditions (3.204) that ϱ is a monic polynomial and that ϱ and σ cannot contain z as a common factor. Moreover it is easy to see that we do not seriously restrict the generality if we assume that both polynomials have no common factor at all; this will be the case from now on.

An important tool for the classification of linear multistep methods is as in the one-step case the *local error.* To define this expression exactly, we consider the linear difference operator $\mathcal{L}$ associated with the difference equation (3.203), which is defined for all continuously differentiable functions φ by

$$\mathcal{L}(\varphi, t, h) := \sum_{j=0}^{k} \Big(\alpha_j \, \varphi(t + jh) - h \, \beta_j \, \varphi'(t + jh) \Big) . \tag{3.206}$$

If we choose in addition the function φ sufficiently smooth, we can expand the right hand side of (3.206) about t and obtain a representation of the form

$$\mathcal{L}(\varphi, t, h) = C_0 \varphi(t) + C_1 \, h \, \varphi'(t) + \cdots + C_\mu \, h^\mu \, \varphi^{(\mu)}(t) + \cdots . \tag{3.207}$$

If there is some $p \in \mathbb{N}$ such that in (3.207)

$$C_\mu = 0 \quad \text{for } \mu = 0, \ldots, p \tag{3.208a}$$

holds, that is, if $\mathcal{L}(\varphi, t, h) = O(h^{p+1})$ for $h \to 0$, then the associated linear multistep method (3.203) is *of (consistency) order p* at least. If in addition

$$C_{p+1} \neq 0 , \tag{3.208b}$$

then the method is exactly of order p. Indeed, it can be shown that the conditions in (3.208) are independent of the special choice of t (cf. J.D.LAMBERT [2]), such that our terminology is well-defined.

Some authors denote $\mathcal{L}(\varphi, t, h)$ as the local error (function), and consequently $\mathcal{L}(y, t_\nu, h)$ as the local error of the method (3.203) at t_ν. In this terminology, (3.208) would mean that the local error of the method is of order $(p+1)$, which fits with the one-step case.

However, there is also another and more intuitive definition of the term local error as follows (cf. e.g. E.HAIRER, S.P.NØRSETT & G.WANNER [1]): Let $y(t)$ be the exact solution of the initial value problem (3.150), and let y_k be obtained from (3.203) by using the exact values $y_0, \dots, y_{k-1}$. Then the local error of the linear multistep method (3.203) is defined as the number

$$y(t_k) - y_k \,. \tag{3.209}$$

Both approaches are more or less equivalent, as can be seen from the following lemma.

Lemma 3.56 (cf. E.HAIRER, S.P.NØRSETT & G.WANNER [1]): *Let $y(t)$ be the exact solution of the ordinary differential equation $y' = f(t,y)$, and y_k be obtained from (3.203) by using the exact values $y(t_0), \dots, y(t_{k-1})$ for $y_0, \dots, y_{k-1}$. Then*

$$y(t_k) - y_k \;=\; \mathcal{L}(y, t_0, h) \cdot \left(1 - h\,\beta_k \frac{\partial}{\partial y} f(t,\eta)\right)^{-1}$$

with some η between $y(t_k)$ and y_k. In particular, we have

$$y(t_k) - y_k \;=\; \mathcal{L}(y, t_0, h)$$

for explicit methods.

Proof. Since the values $y_0, \dots, y_{k-1}$ are assumed to be exact, it follows from (3.203) that y_k is determined implicitly by the equation

$$\sum_{j=0}^{k-1} \Big(\alpha_j\, y(t_j) - h\,\beta_j\, f(t_j, y(t_j))\Big) + y_k - h\,\beta_k f(t_k, y_k) \;=\; 0\,,$$

or, in terms of the the operator $\mathcal{L}$:

$$\mathcal{L}(y, t_0, h) + (y_k - y(t_k)) - h\,\beta_k \Big(f(t_k, y_k) - f(t_k, y(t_k))\Big) \;=\; 0\,.$$

Here we have used that $f(t_j, y(t_j)) = y'(t_j)$. Since, by the mean value theorem, there exists a number η between $y(t_k)$ and y_k, such that

$$f(t_k, y_k) - f(t_k, y(t_k)) \;=\; \frac{\partial}{\partial y} f(t,\eta) \cdot (y_k - y(t_k))\,,$$

the result follows. □

There is also a very nice algebraic characterization of the consistency order in terms of the characteristic polynomials ϱ and σ and their coefficients:

Theorem 3.57: *Let $p \in \mathbb{N}$. The linear multistep method (3.203) is of order p, if and only if*

$$\sum_{j=0}^{k} \alpha_j j^{\mu} = \mu \sum_{j=0}^{k} \beta_j j^{\mu-1} \tag{3.210a}$$

holds for $\mu = 0, \ldots, p$. The order is exactly equal to p, if in addition

$$\sum_{j=0}^{k} \alpha_j j^{p+1} \neq (p+1) \sum_{j=0}^{k} \beta_j j^{p}. \tag{3.210b}$$

For $p = 1$, the conditions in (3.210a) take the form

$$\varrho(1) = 0 \quad \textit{and} \quad \varrho'(1) = \sigma(1) \tag{3.211}$$

and are usually denoted as consistency conditions.

Proof. After some straightforward calculations we obtain the following formulae for the coefficients C_μ in (3.207):

$$C_0 = \sum_{j=0}^{k} \alpha_j,$$

$$C_\mu = \frac{1}{\mu!} \sum_{j=0}^{k} \alpha_j j^{\mu} - \frac{1}{(\mu-1)!} \sum_{j=0}^{k} \beta_j j^{\mu-1} \quad \text{for } \mu \geq 1.$$

This already implies the equivalence of (3.210) with (3.208), while the last statement is obvious. □

It is already indicated by the conditions (3.211) that the characteristic polynomials ϱ and σ, in particular the zeros of ϱ, play an important role in the consistency theory of linear multistep methods; this comes out even more clearly within the concept of stability, as worked out by G. Dahlquist in a series of papers (G.DAHLQUIST [1,2,3], see also G.DAHLQUIST [4]). The key definition of this concept is as follows.

Definition 3.58: Let $\varrho(z)$ again denote the first characteristic polynomial of the linear multistep method (3.203).
(i) The method is called *stable*, if the zeros of ϱ are all on or within the unit circle, and if those on the circle have multiplicity 1.
(ii) The method is called *strictly stable*, if in addition $z_1 := 1$ is the only zero of ϱ on the circle.

Note that, according to (3.211), $z_1 := 1$ is a zero for each consistent method; so, in a certain sense, strict stability is the best we can get.

The motivation for Definition 3.58 comes from the following theorem, which is well-known within the theory of linear difference equations, e.g. P.HENRICI [1] or G.MEINARDUS & G.MERZ [2].

Theorem 3.59: *Consider the homogeneous problem for the linear multistep method (3.203), that is,*

$$\sum_{j=0}^{k} \alpha_j \, y_{\nu+j} \; = \; 0 \,, \tag{3.212}$$

and denote by $z_1, \ldots, z_s$ the zeros of the characteristic polynomial $\varrho(z)$ with corresponding multiplicities $\lambda_1, \ldots, \lambda_s$. Then the general solution of (3.212) is of the form

$$y_\nu \; = \; \sum_{j=1}^{s} z_j^\nu \sum_{\mu=0}^{\lambda_j - 1} \gamma_{\mu j} \, \nu^\mu \,. \tag{3.213}$$

From (3.213) it is clear that the sequence of solutions $\{y_\nu\}$ remains bounded if and only if the method is stable in the sense of Definition 3.58.

In the context of Runge-Kutta methods we had encountered the Butcher barries, which told us about the attainable order for an $s-$ stage Runge-Kutta method, see Theorem 3.50. Here, in the context of linear multistep methods, there exist similar results, the so-called *(first) Dahlquist barriers,* concerning the attainable order for a stable linear $k-$ step method. As in the Runge-Kutta case, we refer for a proof of the following deep result to the original papers or to textbooks on ordinary differential equations.

Theorem 3.60 (G.DAHLQUIST [1,2]): *Let*

$$p_{max}(k) \;:=\; \max\{p \in \mathbb{N} \mid \text{there exists a stable linear } k\text{–step method of order } p\}\;.$$

Then we have

$$p_{max}(k) \;=\; \begin{cases} k+2\,, & \text{if } k \text{ is even and } \beta_k > 0, \\ k+1\,, & \text{if } k \text{ is odd and } \beta_k > 0, \\ k\,, & \text{if } \beta_k \le 0. \end{cases}$$

In particular, $p_{max}(k) = k$ *for explicit* $k-$ *step methods.*

A nice generalization of these Dahlquist barriers for multistep-multiderivative methods can be found in M.REIMER [1].

It is about time for some examples. We begin with the methods of *Adams* type, which are of the form

$$y_{\nu+k} - y_{\nu+k-1} \;=\; h \cdot \sum_{j=0}^{k} \beta_j\, f_{\nu+j}\;; \tag{3.214}$$

if they are explicit, one speaks of *Adams-Bashforth* methods, while the implicit ones are known as *Adams-Moulton* methods.

Since in every case the first characteristic polynomial equals

$$\varrho(z) \;=\; z^k - z^{k-1}$$

with the simple zero $z_1 = 1$ and the $(k-1)-$ fold one $z_2 = 0$, all Adams methods are strictly stable. Furthermore it can easily be shown (cf. Problem 3.19) that for each k there exists an explicit $k-$ step Adams method of order k, and also an implicit one of order $(k+1)$. So, although quite simple in shape, Adams methods provide efficient numerical algorithms.

A second important subclass of linear multistep methods is defined through the characteristic polynomial

$$\varrho(z) = z^k - z^{k-2} .$$

The explicit members of this subclass are denoted as *Nyström* methods, the implicit ones as *generalized Milne-Simpson* rules. Since the characteristic polynomial ϱ has now an additional simple zero at -1, these methods are stable, but not strictly stable. Among the most efficient Nyström methods is the midpoint rule

$$y_{\nu+2} - y_\nu = 2hf_{\nu+1} , \tag{3.215}$$

an explicit 2-step method, which plays a key role in the famous method of Gragg, see Subsection 4.3.2.

A quite large class of methods, which are stable, and in some cases even strictly stable, is that of methods whose characteristic polynomial is a *Cauchy polynomial.* This important fact seems to have so far not drawn much attention on itself, and we therefore take us the time (and about 3 pages of this book) to work it out in some detail.

Let us for the moment write the characteristic polynomial ϱ in the more convenient form

$$\varrho(z) = z^k - \sum_{j=1}^{k} d_j z^{k-j} . \tag{3.216}$$

We denote ϱ as a *Cauchy polynomial*, if the coefficients $d_j(= -\alpha_{k-j})$ satisfy

$$\begin{aligned} \sum_{j=1}^{k} d_j &= 1, \quad \text{and} \\ 0 \le d_j \le 1 & \quad \text{for } j = 1, \ldots, m . \end{aligned} \tag{3.217}$$

Theorem 3.61 (K.P.HADELER & G.MEINARDUS [1], G.MEINARDUS & G.WALZ [1]): *Suppose that the first characteristic polynomial ϱ of a linear multistep method is a Cauchy polynomial. Then the method is stable.*

Furthermore, let be

$$d_{\mu_1}, \ldots, d_{\mu_s}$$

the non-zero coefficients in (3.216), and denote by g the greatest common divisor of the subscripts $\mu_1, \ldots, \mu_s$. Then the zeros of ϱ on the unit circle are just the gth roots of unity, i.e. the numbers

$$e^{2\pi i\nu/g}, \quad \nu = 0, \ldots, g-1 .$$

In particular, if $g = 1$, then the method is strictly stable.

Remark. The characteristic polynomials of the Adams as well as the Nyström type methods are Cauchy polynomials; in the terminology of Theorem 3.61 we have $g = 1$ for Adams and $g = 2$ for Nyström type methods, which proves again the above-stated assertions.

Proof of Theorem 3.61. Let $\{y_\nu\}$ be the general solution of the homogeneous problem

$$y_{\nu+k} - \sum_{j=1}^{k} d_j \, y_{\nu+k-j} = 0 .$$

Because of Conditions (3.217) we obtain for $\nu \in \mathbb{N}_0$ the estimation

$$|y_{\nu+k}| \leq \sum_{j=1}^{k} d_j \, |y_{\nu+k-j}| \leq \max_{\mu=\nu,\dots,\nu+k-1} \{|y_\mu|\} .$$

Hence, for $\nu \in \mathbb{N}_0$,

$$|y_{\nu+k}| \leq \max_{\mu=0,\dots,k-1} \{|y_\mu|\} ,$$

which proves that the sequence $\{y_\nu\}$ remains bounded. Using Theorem 3.59 we recognize that the method is stable.

To prove the second assertion, we first note that there exist natural numbers

$$\kappa_1, \kappa_2, \dots, \kappa_s$$

with greatest common divisor 1 such that

$$\mu_\rho = g \cdot \kappa_\rho , \quad \rho = 1, 2, \dots, s .$$

We assume that the subscripts μ_ρ satisfy the relation

$$1 \leq \mu_1 \leq \dots \leq \mu_s \leq k .$$

Then, with the complex variable $w := z^g$, we get

$$\varrho(z) = z^{k-\mu_s} \cdot \Big(w^{\kappa_s} - \sum_{\rho=1}^{s} d_{\mu_\rho} w^{\kappa_s - \kappa_\rho} \Big) =: z^{k-\mu_s} \cdot \tilde{\varrho}(w) ,$$

where $\tilde{\varrho}$ is a (normalized) Cauchy polynomial of degree κ_s . Now with

$$\zeta = \mathrm{e}^{2\pi i \nu / g} , \quad \nu = 0, 1, \dots, g-1 ,$$

it follows

$$\varrho(\zeta) = \zeta^{k-\mu_s} \cdot \tilde{\varrho}(\zeta^g) = \zeta^{k-\mu_s} \cdot \tilde{\varrho}(1) = 0 .$$

Conversely, let ζ be a zero of $\varrho(z)$ of absolute value 1. Then, putting $w_0 = \zeta^g$, we get

$$d_{\mu_1} w_0^{-\kappa_1} + d_{\mu_2} w_0^{-\kappa_2} + \ldots + d_{\mu_s} w_0^{-\kappa_s} = 1 .$$

This convex combination can only have the value 1, if simultaneously the relations

$$w_0^{\kappa_1} = w_0^{\kappa_2} = \ldots = w_0^{\kappa_s} = 1$$

hold. Since the natural numbers $\kappa_1, \kappa_2, \ldots, \kappa_s$ have the greatest common divisor 1, there exist integers $i_1, i_2, \ldots, i_s$ such that

$$i_1\kappa_1 + i_2\kappa_2 + \ldots + i_s\kappa_s = 1 .$$

Hence

$$\begin{aligned} w_0 &= w_0^{i_1\kappa_1 + i_2\kappa_2 + \ldots + i_s\kappa_s} \\ &= (w_0^{\kappa_1})^{i_1} (w_0^{\kappa_2})^{i_2} \cdots (w_0^{\kappa_s})^{i_s} = 1 , \end{aligned}$$

i.e. $\zeta^g = 1$ and thus

$$\zeta = e^{2\pi i\nu/g}$$

with some $\nu \in \mathbb{Z}$. □

3.6.4 Existence of Asymptotic Expansions. Let us come now to a major topic of this section, the existence of an asymptotic expansion for the final result of a multistep method. In the case of linear methods, which we consider mainly, this was shown the first time with a long and difficult proof by W. B. Gragg in his famous thesis (W.B.GRAGG [1], cf. also H.J.STETTER [3]).

However, sometimes things seem to become easier to understand, if you make them a little bit more complicated (in the sense of more general) first. This is indeed the case for the present situation, since it had turned out to be useful to write linear multistep methods in a more general context as one-step methods in a higher dimensional space and to prove existence of an asymptotic expansion in that space. This will be worked out on the following pages.

Building on earlier work of J.C.BUTCHER [2] and R.SKEEL [1], the approach we are going to present here is essentially due to E.HAIRER & CH.LUBICH [1]; in particular, they proved by generalizing the proof of the one-step case (Theorem 3.46) that the final result of a quite general class of fixed stepsize methods possesses a logarithmic asymptotic expansion.

We consider again the initial value problem (3.150) and assume that f satisfies the usual regularity conditions, essentially that it is sufficiently smooth in a suitable region in $\mathbb{R}^2$. The methods we will consider from now on are defined as follows:

Definition 3.62: A *general (fixed stepsize) method* is a procedure consisting of the following three parts:
(i) A so-called *forward step procedure*; this is a mapping from $\mathbb{R}^k$ into itself of the form

$$u_{\nu+1} = Su_\nu + h\,\Phi(t_\nu, u_\nu, h)\ ,\ \nu = 0, 1, \dots\ , \tag{3.218a}$$

with a square matrix S, independent of the problem, and an increment function Φ.
(ii) A *starting procedure* $\varphi = \varphi(h)$, which is assumed to be sufficiently smooth. The starting procedure determines the starting value u_0 of (3.218), i.e.

$$u_0 = \varphi(h)\ . \tag{3.218b}$$

(iii) A *correct value function* $z = z(t,h)$, which must also be smooth enough. The idea is that the quantities $z_\nu := z(t_\nu, h)$, which give the desired correct value, are approximated by the corresponding values u_ν, such that the global error is given by $u_n - z_n := u(x,h) - z(x,h)$. It is tacitly assumed that the exact solution y of (3.150) can be recovered from z.

Let us see how linear multistep methods of the standard form (3.203) fit into this framework: We define the matrix S as

$$S := \begin{pmatrix} -\alpha_{k-1} & -\alpha_{k-2} & \cdots & \cdots & -\alpha_0 \\ 1 & 0 & \cdots & \cdots & 0 \\ 0 & 1 & 0 & \cdots & 0 \\ \vdots & & \ddots & \ddots & \vdots \\ 0 & \cdots & 0 & 1 & 0 \end{pmatrix}, \tag{3.219}$$

and the increment function Φ by

$$\Phi(t,u,h) := (1,0,\ldots,0)^T \cdot \tilde{\Phi}(t,u,h) , \tag{3.220}$$

where, for any vector $u = (u^{k-1},\ldots,u^0)$, $\tilde{\Phi} = \tilde{\Phi}(t,u,h)$ is implicitly defined through the equation

$$\tilde{\Phi} := \sum_{j=0}^{k-1} \beta_j\, f(t+jh,u^j) + \beta_k \big(f(t+kh, h\tilde{\Phi} - \sum_{j=0}^{k-1} \alpha_j u^j)\big) . \tag{3.221}$$

(Note that, with $\tilde{\Phi}$ defined by (3.221), the linear multistep method (3.203) kann be written in the form

$$y_{\nu+k} = -\sum_{j=0}^{k-1} \alpha_j y_{\nu+j} + h\,\tilde{\Phi}(t_\nu, (y_\nu,\ldots,y_{\nu+k-1}), h)\ ) .$$

As starting procedure φ, we might take any of the one-step methods already studied, as long as it is sufficiently smooth, such that it possesses an asymptotic expansion of the desired type; φ then determines the starting values $\varphi(h) = (y_{k-1},\ldots,y_0)$. Finally, a suitable correct value function z is given by

$$z(t,h) = \big(y(t+(k-1)h),\ldots,y(t)\big) ,$$

where y denotes the exact solution of the problem.

Clearly, there are a lot of other methods which can be considered as general methods in the sense of Definition 3.62, e.g. classical one-step methods or predictor-corrector-methods; for more information on these topics, see E.HAIRER & CH.LUBICH [1] or R.SKEEL [1].

We still have to introduce some notations and definitions concerning the general method (3.218). The method is called *stable*, if the sequence of matrices $\{S^n\}_{n\in\mathbb{N}}$ is uniformly bounded for $n \to \infty$. Moreover, it is called *strictly stable*, if

$$E := \lim_{n\to\infty} S^n \tag{3.222}$$

exists. Note that, since the characteristic polynomial ϱ of the linear multistep method (3.203) coincides with the characteristic polynomial of the matrix S in (3.219), the above definition of stability is, in the case of linear multistep methods, equivalent to Definition 3.58.

The *local error* d_ν of Method (3.218) is defined as

$$\begin{aligned} d_0 &:= z_0 - \varphi(h)\,, \quad \text{and} \\ d_\nu &:= z_\nu - S z_{\nu-1} - h\,\Phi(t_{\nu-1}, z_{\nu-1}, h)\,, \quad \nu = 1, 2, \ldots . \end{aligned} \tag{3.223}$$

This is in complete analogy to the one-step case (Eqn. (3.156)) and to (3.209), and therefore, according to Lemma 3.56, for linear multistep methods equivalent to our local error functional $\mathcal{L}$.

Finally, we have to introduce the term (consistency) order of the general method (3.218). Since we always assume that the several functions involved (in particular the procedures φ, Φ and the correct value function z) are sufficiently smooth, we can expand the local error into a Taylor series in h, i.e.

$$\begin{aligned} d_0 &= \sum_{\mu=0}^{p-1} \gamma_\mu h^\mu + O(h^p)\,, \quad \text{and} \\ d_{\nu+1} &= \sum_{\mu=0}^{p} d_\mu(t_\nu) h^\mu + O(h^{p+1})\,. \end{aligned} \tag{3.224}$$

We say that the method is *(consistent) of order* p, if it is strictly stable and if the following conditions hold:

$$\begin{aligned} d_\nu &= O(h^p) \quad \text{for } h \to 0 \text{ and } \nu \le n\,, \text{ and} \\ Ed_p(t) &= 0 \end{aligned} \tag{3.225}$$

with $d_p(t)$ from (3.224) and the limit matrix E from (3.222) (remember that h and n are coupled by the assumption $h = (x - t_0)/n$, i.e. we have $n \cdot h =$ constant).

Unfortunately, conditions (3.225) are quite technical in nature and do maybe not allow an easy insight, but it can be shown (cf. E.HAIRER, S.P.NØRSETT & G.WANNER [1]) that they are equivalent to the easier-to-understand conditions

$$d_\nu \; = \; O(h^p) \quad \text{and} \quad Ed_{\nu+1} \; = \; O(h^{p+1})$$

for $h \to 0$ and $\nu \le n$. The connection to linear multistep methods is now obvious: There, d_ν denotes the usual local error, while $Ed_{\nu+1}$ is just the local error divided by $\beta_0 + \cdots + \beta_k = \sigma(1)$.

We are now prepared to prove the first main theorem of this subsection, which states the existence of an logarithmic asymptotic expansion for strictly stable methods of the type (3.218):

Theorem 3.63 (E.HAIRER & CH.LUBICH [1]): *Consider a general method as defined in Definition 3.62, which is strictly stable and consistent of order* p, $p \in \mathbb{N}$. *Then, for all* t_ν, *the error of this method possesses an asymptotic expansion of the form*

$$u_\nu \;=\; z_\nu + \sum_{\mu=p}^{m} e_\mu(t_\nu)\, h^\mu + R(t_\nu, h)\, h^{m+1} \tag{3.226a}$$

with certain coefficient functions e_μ; *the order* m *of this expansion depends on the smoothness of the several functions involved. Here,* $R(t,h)$ *is a remainder term which is bounded uniformly for* $h \to 0$ *as* $t \ge t_0 + \delta$, $\delta > 0$ *fixed.*

In particular, for the global error

$$u_n \;=\; z_n + \sum_{\mu=p}^{m} e_\mu(x)\, h^\mu + R(x,h)\, h^{m+1}\,. \tag{3.226b}$$

Remark. Theorem 3.63 indicates that we have here something like a "pseudo-logarithmic" asymptotic expansion, since the remainder $R(t,h)$ is not bounded for $t \to t_0$ (which means that we cannot make use of our beloved "O"-notation here). To make this even more precise, we shall prove the existence of an expansion of the form

$$u_\nu \;=\; z_\nu + \sum_{\mu=p}^{m} \left(e_\mu(t_\nu) + \varepsilon_\nu^\mu\right) h^\mu + R_1(h)\, h^{m+1}\,, \tag{3.227}$$

where for all μ

$$\varepsilon_\nu^\mu \;=\; O(\varrho^\nu) \qquad \text{for } \nu \to \infty$$

with some $\varrho < 1$ (see proof) and $R_1(h)$ bounded for $h \to 0$.

Representations (3.226a) and (3.227) are connected by the relation

$$R(t_\nu,h) \;=\; R_1(h) + \sum_{\mu=1}^{p-m-1} \varepsilon_n^{m+1-\mu}\, h^{-\mu}\,, \tag{3.228}$$

from which we recognize that the remainder term $R(x,h)$ may be not be bounded for $h \to 0$ as x comes too close to t_0. However, if x is bounded away from t_0, say $x \geq t_0 + \delta$, $\delta > 0$, then each of the sequences $\{\varepsilon_\nu^\mu\}$ converges to zero faster than any power of $\delta/\nu \leq h$.

Proof of Theorem 3.63. We shall imitate the proof of Theorem 3.46 (the one-step case) and use induction with respect to μ. Before coming to that we remark that this inductive proof will force us to treat increment functions of the form

$$\Phi_\nu(t,u,h) \;=\; \Phi(t, u + h a_\nu(h), h) + b_\nu(h) \tag{3.229}$$

with sufficiently smooth Φ and polynomials a_ν, b_ν, whose coefficients, say a_ν^j and b_ν^j, satisfy for all j the relation

$$a_\nu^j = O(\rho^\nu) \quad \text{and} \quad b_\nu^j = O(\rho^\nu) \qquad \text{for } \nu \to \infty .$$

Here, ρ is some number with $\rho(S-E) < \rho < 1$; this inclusion is possible since the spectral radii of S and E satisfy $\rho(S) = 1$ and $\rho(E) > 0$, due to the fact that 1 is an eigenvalue of S, such that the sequence $\{S^n\}_{n\in\mathbb{N}}$ cannot converge to zero. Moreover, using Taylor expansion we see that the local error of Method (3.229) possesses an asymptotic expansion in powers of h in analogy to (3.224), but now with a perturbation term δ_ν^μ, satisfying $\delta_\nu^\mu = O(\rho^\nu)$ for $\nu \to \infty$, added:

$$\begin{aligned} d_0 &= \sum_{\mu=0}^{m} \gamma_\mu h^\mu + O(h^{m+1}), \quad \text{and} \\ d_{\nu+1} &= \sum_{\mu=0}^{m+1} (d_\mu(t_\nu) + \delta_\nu^\mu) h^\mu + O(h^{m+2}) . \end{aligned} \tag{3.230}$$

Now let us start with the main part of the induction and put $\mu = p$. Similar as in the proof of Theorem 3.46 we consider a new increment function $\Phi_\nu^{(1)}$ and corresponding starting procedure $\varphi^{(1)}$ of the form

$$\begin{aligned} \varphi^{(1)}(h) &= \varphi(h) - (e(t_0) - \varepsilon_0)\, h^p , \quad \text{and} \\ \Phi_\nu^{(1)}(t,u,h) &= \Phi_\nu(t, u + (e(t) + \varepsilon_\nu) h^p, h) \\ &\qquad - (e(t+h) - Se(t)) h^{p-1} \\ &\qquad\qquad - (\varepsilon_{\nu+1} - S\varepsilon_\nu) h^{p-1} , \end{aligned} \tag{3.231}$$

where $e(t)$ and $\{\varepsilon_\nu\}$ still have to be determined.

These procedures define the new method

$$\begin{aligned} u_0^{(1)} &= \varphi^{(1)}(h) , \quad \text{and} \\ u_{\nu+1}^{(1)} &= Su_\nu^{(1)} + h\, \Phi_\nu^{(1)}(t_\nu, u_\nu^{(1)}, h) \end{aligned} \tag{3.232}$$

with the numerical solution

$$u_\nu^{(1)} = u_\nu - (e(t_\nu) - \varepsilon_\nu)h^p .$$

Since Φ_ν was assumed to be of the form (3.229), the same holds true for $\Phi_\nu^{(1)}$, such that its local error also possesses an asymptotic expansion like (3.230).

Next we determine $e(t)$ and ε in such a way that the method (3.232) is consistent of order $(p+1)$. To do this we expand the local error $d_\nu^{(1)}$ of this method and obtain

$$\begin{aligned} d_0^{(1)} &= z_0 - u_0^{(1)} = (\gamma_p + e(t_0) + \varepsilon_0)h^p + O(h^{p+1}) , \text{ and} \\ d_{\nu+1}^{(1)} &= z_{\nu+1} - Sz_\nu - h\Phi_\nu^{(1)}(t_\nu, z_\nu, h) \qquad (3.233) \\ &= d_{\nu+1} + \Big((I-S)e(t_\nu) + (\varepsilon_{\nu+1} - S\varepsilon_\nu)\Big)h^p \\ &\quad + \Big(-G(t_\nu)(e(t_\nu) + \varepsilon_\nu) + e'(t_\nu)\Big)h^{p+1} + O(h^{p+2}) , \end{aligned}$$

where

$$G(t) = \frac{\partial}{\partial u}\Phi_\nu^{(1)}(t, z(t,0), 0)$$

is independent of ν by (3.229), since $\beta_\nu(0) = 0$, due to the consistency.

From (3.233) and (3.225) we deduce that the new method is consistent of order $(p+1)$, if the following relations hold:

$$\begin{aligned} (i)\quad & \varepsilon_0 = -\gamma_p - e(t_0) , \\ (ii)\quad & d_p(t) + (I-S)e(t) + \delta_\nu^p + \varepsilon_{\nu+1} - S\varepsilon_\nu = 0 , \quad \text{and} \\ (iii)\quad & Ee'(t) = EG(t)e(t) - Ed_{p+1}(t) . \end{aligned}$$

To prove the solvability of this system, we split it up once more and remark that condition (ii) is satisfied if there exists a constant vector c such that

$$\begin{aligned} (iia)\quad & d_p(t) = (S-I)(e(t)+c) , \quad \text{and} \\ (iib)\quad & \varepsilon_{\nu+1} - c = S(\varepsilon_\nu - c) - \delta_\nu^p \end{aligned}$$

hold simultaneously.

Using the equality $(I-S+E)^{-1}(I-S)=(I-E)$, we recognize that condition (iia) is equivalent to

$$(I-S+E)^{-1}d_p(t) \;=\; -(I-E)(e(t)+c)\,. \tag{3.234}$$

Note that here and below we make several times use of the relation

$$S^i\cdot E^j \;=\; E \quad \text{for all } i\in \mathbb{N}_0 \text{ and } j\in\mathbb{N}\,. \tag{3.235}$$

From condition (i) we obtain $\varepsilon_0-c=-\gamma_p-(e(t_0)+c)$, such that by (3.234)

$$(I-E)(\varepsilon_0-c) \;=\; -(I-E)\gamma_p+(I-S+E)^{-1}d_p(t_0)\,. \tag{3.236}$$

For (3.236) to hold it is sufficient that

$$\varepsilon_0-c \;=\; -(I-E)\gamma_p+(I-S+E)^{-1}d_p(t_0)\,, \tag{3.237}$$

since $Ed_p(t_0)=0$, due to the consistency of the old method.

Now the numbers $\varepsilon_\nu-c$ are determined by the recurrence relation (iib), which yields

$$\begin{aligned}\varepsilon_\nu-c \;&=\; S^\nu(\varepsilon_0-c)-\sum_{j=1}^{\nu}S^{\nu-j}\delta^p_{j-1}\\ &=\; E(\varepsilon_0-c)+(S-E)^\nu(\varepsilon_0-c)-E\sum_{j=0}^{\infty}\delta^p_j\\ &\qquad +E\sum_{j=\nu}^{\infty}\delta^p_j-\sum_{j=1}^{\nu}(S-E)^{\nu-j}\delta^p_{j-1}\,,\end{aligned}$$

since $S^\nu=E+(S-E)^\nu$, due to (3.235). Putting now

$$c \;:=\; E\cdot\sum_{j=0}^{\infty}\delta^p_j\,, \tag{3.238}$$

and choosing some ρ with $1 > \rho > \rho(S - E)$, the sequence of these ε_ν 's satisfies

$$\varepsilon_\nu = O(\rho^\nu) \quad \text{for } \nu \to \infty ,$$

since $E(\varepsilon_0 - c) = 0$ by (3.237) and because $\delta_\nu^p = O(\rho^\nu)$.

In order to determine the function $e(t)$ we set $v(t) := Ee(t)$. Since Equation (3.238) implies $(I - E)c = 0$, we see using (3.234) that $e(t)$ can be recovered from $v(t)$ by the relation

$$e(t) = v(t) - (I - S + E)^{-1} d_p(t) . \tag{3.239}$$

Condition (iii) can now be interpreted as differential equation for $v(t)$,

$$v'(t) = EG(t)\Big(v(t) - (I - S + E)^{-1} d_p(t)\Big) - Ed_{p+1}(t) \tag{3.240a}$$

and condition (i) yields the starting value

$$v(t_0) = -E(\gamma_p + \varepsilon_0) . \tag{3.240b}$$

The initial value problem (3.240) can now be solved to determine $v(t)$, and we recover $e(t)$ by (3.239). Altogether, we have thus proved the solvability of the system $(i) - (iii)$ and therefore the consistency of order $p + 1$ of the new method $u_\nu^{(1)}$.

The rest of the proof now follows at once, since we can use a result of Skeel (R.Skeel [1], Theorem 3.8, see also Theorem 8.13 in E.Hairer, S.P.Nørsett & G.Wanner [1]), which says that in complete analogy to the case of classical linear multistep methods, also in the theory of general methods strict stability and consistency of order $p+1$ imply convergence of order $p+1$. Therefore we have proved that

$$u_\nu^{(1)} - z_\nu = O(h^{p+1}) \quad \text{for } \nu \le n \text{ and } h \to 0 ,$$

which settles the existence of the first term of the desired expansion.

If $m = p$, the proof is complete. Otherwise, we repeat the procedure from above, thus defining new increment functions $\Phi^{(2)}, \Phi^{(3)}, \ldots$, until the existence of the full asymptotic expansion is established. □

So far, we have proved the existence of an asymptotic expansion for strictly stable methods. However, sometimes strict stability is too strict a restriction; for example, all Nyström type methods, in particular the midpoint rule (3.215), are not strictly stable. We shall therefore extend and modify the proof of Theorem 3.63 to cover stable (and not necessarily strictly stable) methods. This is the case when the sequence $\{S^n\}_{n\in\mathbb{N}}$ is bounded, but not necessarily converging, or in other words when the minimal polynomial of the matrix S has several (simple) roots, say $z_1, \ldots, z_l$, on the unit circle. Note that again linear multistep methods which are stable in the sense of Definition 3.58 fall into this class.

We shall restrict ourselves to the case when $z_1, \ldots, z_l$ are certain roots of unity. This is by far the most important situation, covering for example Nyström and generalized Milne-Simpson rules and all methods whose first characteristic polynomial is of Cauchy type, see Theorem 3.61. For more information on the general case we refer to the original paper by E.HAIRER & CH.LUBICH [1].

Theorem 3.64 (E.HAIRER & CH.LUBICH [1]): *Suppose that the general method (3.218) is stable and consistent of order p, $p \in \mathbb{N}$. Moreover, assume that the zeros $z_1, z_2, \ldots, z_l$ (with $z_1 = 1$) of the minimal polynomial of S satisfy $z_j^q = 1$ for $j = 1, \ldots, l$ and some $q \in \mathbb{N}$. Then, with $\zeta := e^{2\pi i/q}$, the error of this method satisfies an asymptotic relation of the form*

$$u_\nu = z_\nu + \sum_{s=0}^{q-1} \zeta^{\nu s} \Big(\sum_{\mu=p}^{m} e_{\mu s}(t_\nu)\, h^\mu \Big) + r(\nu)\, h^{m+1}, \qquad (3.241)$$

where r is uniformly bounded for $\delta_1 \le \nu h \le$ constant. In particular, Relation (3.241) holds for $\nu = n$, i.e., for the global error.

Remark. For the special case of the midpoint rule, where $k, l, q = 2$, this result is due to W.B.GRAGG [1], see also H.J.STETTER [2,3] and R.BULIRSCH & J.STOER [2]. Moreover, we want to point out that (3.241) does not establish an asymptotic expansion in the sharp sense, since the $\zeta^{\nu s}$ depend on ν; see also the remarks following Theorem 4.15.

Proof of Theorem 3.64. Let us first see how the matrix S behaves in this case. Using the Jordan canonical form, say $S = T \cdot J \cdot T^{-1}$, S can be decomposed into the form

$$S = \sum_{j=1}^{l} z_j E_j + R,$$

where each E_j is a block diagonal matrix of the form

$$E_j = T \cdot \operatorname{diag} \{\underbrace{0, \ldots, 0, I, 0, \ldots, 0}_{(l+1)\text{ blocks}}\} \cdot T^{-1}$$

with I standing in the j th (block) component, and

$$R = T \cdot \operatorname{diag} \{0, \ldots\ldots, 0, J\} \cdot T^{-1}$$

with $\rho(R) < 1$.

The very nice idea of the proof, due to E.HAIRER & CH.LUBICH [1], is to consider q consecutive steps of the method (3.218) as one large step. Using, with some integer i, $0 \le i \le q-1$, the abbreviations $u_\nu^* = u_{\nu q+i}$, $h^* = qh$ and $t_\nu^* = t_i + \nu h^*$, the method can be rewritten as

$$u_{\nu+1}^* = S^q u_\nu^* + h^* \Phi^*(t_\nu^*, u_\nu^*, h^*) \tag{3.242}$$

with some suitable Φ^*. Due to the above-mentioned structure of the matrix S, we have for all $\kappa \in I\!N_0$ the relation

$$S^\kappa = \sum_{j=1}^{l} z_j^\kappa E_j + R^\kappa \tag{3.243}$$

and in particular

$$S^q = \sum_{j=1}^{l} E_j + R^q ,$$

which implies that

$$\lim_{\nu\to\infty} (S^q)^\nu = \sum_{j=1}^{l} E_j := E^*$$

exists. Therefore, Method (3.242) is strictly stable, and we may apply Theorem 3.63, if we can prove the consistency of the method. To do this we remark that, after some straightforward calculations, the local error of (3.242) turns out to satisfy

$$\begin{aligned} d_0^* &= O(h^p), \quad \text{and} \\ d_{\nu+1}^* &= (I + S + \cdots + S^{q-1})\, d_p(t_\nu^*) h^p + O(h^{p+1}) . \end{aligned} \tag{3.244}$$

From (3.243) we obtain, using $I = E^* + R^0$,

$$\begin{aligned} \sum_{\kappa=0}^{q-1} S^\kappa &= \sum_{\kappa=0}^{q-1} \Big(\sum_{j=1}^{l} z_j^\kappa E_j + R^\kappa \Big) \\ &= I + \sum_{j=1}^{l} E_j \sum_{\kappa=1}^{q-1} z_j^\kappa + \sum_{\kappa=1}^{q-1} R^\kappa \\ &= I + qE_1 - E^* + \sum_{j=2}^{l} E_j \frac{1 - z_j^q}{1 - z_j} + \sum_{\kappa=1}^{q-1} R^\kappa \\ &= I + qE_1 - E^* + \sum_{\kappa=1}^{q-1} R^\kappa . \end{aligned}$$

Inserting this into (3.244), it turns out that

$$E^*(I + S + \cdots + S^{q-1})\, d_p(t) \;=\; qE_1 d_p(t) \;=\; 0\,,$$

due to (3.225). Hence, Method (3.242) is of consistent of order p.

Application of Theorem 3.63 to this method now yields

$$u_{\nu q+i} \;=\; z_{\nu q+i} + \sum_{\mu=p}^{m} e^*_{\mu i}(t_{\nu q+i})\, h^\mu + r_i(\nu) h^{m+1}\,,$$

where r_i possesses the desired boundedness-properties. Now we define, for $\mu = p, \ldots, m$, $e_{\mu s}(t)$ as solution of the Vandermonde type system

$$\sum_{s=0}^{q-1} \zeta^{is} e_{\mu s}(t) \;=\; e^*_{\mu i}(t)\,,$$

which completes the proof of Theorem 3.64. □

In analogy to the one-step case, there is great interest in general methods possessing an h^2- expansion; such expansions are usually provided by symmetric methods, which we will study now shortly. First we have to find a suitable definition of the term "symmetry"; to do this, it is (once more in analogy to the one-step case, cf. (3.170)) useful to rewrite Method (3.218) in the form

$$\begin{aligned} u(t_0, h) &= \varphi(h) \quad \text{and} \\ u(t+h, h) &= S\, u(t,h) + h\, \Phi(t, u(t,h), h)\,, \qquad (3.245) \\ &\qquad t = t_0 + \nu h, \nu \in \mathbb{Z}\,. \end{aligned}$$

Definition 3.65: Assume that the correct value function z of our method (3.245) satisfies the symmetry relation

$$z(t,h) \;=\; Q\, z(t+qh, -h) \qquad (3.246)$$

with a square matrix Q and an integer q.

Then the method (3.245) is called *symmetric,* if the numerical solution u satisfies the analogue of (3.246), i.e. if

$$u(t,h) \;=\; Q\,u(t+qh,-h)\,. \tag{3.247}$$

For linear multistep methods the correct value function reads

$$z(t,h) \;=\; \big(y(t+(k-1)h),\ldots,y(t)\big)\,,$$

hence (3.246) holds with $q=k-1$ and

$$Q \;=\; \begin{pmatrix} 0 & \cdots & \cdots & 0 & 1 \\ \vdots & & \cdot^{\cdot^{\cdot}} & 1 & 0 \\ \vdots & \cdot^{\cdot^{\cdot}} & \cdot^{\cdot^{\cdot}} & \cdot^{\cdot^{\cdot}} & \vdots \\ 0 & \cdot^{\cdot^{\cdot}} & \cdot^{\cdot^{\cdot}} & & \vdots \\ 1 & 0 & \cdots & \cdots & 0 \end{pmatrix},$$

where of course first h has to be replaced by $-h$, and then t shifted to $t+qh$.

If the coefficients of the linear multistep method satisfy

$$\alpha_{k-j} = -\alpha_j \quad \text{and} \quad \beta_{k-j} = \beta_j \quad \text{for all } j \tag{3.248a}$$

or equivalently, if

$$\varrho(z) = -z^k\,\varrho(z^{-1}) \quad \text{and} \quad \sigma(z) = z^k\,\sigma(z^{-1})\,, \tag{3.248b}$$

then, after some lengthy but straightforward calculations one recognizes that (3.247) is satisfied for all $t = t_0 + \nu h$ if and only if it is satisfied for $t = t_0$. This means that a linear multistep method is symmetric in the sense of Definition 3.65 if it is so in the classical sense (i.e. if (3.248) holds) and if the starting procedure $\varphi(h)$ satisfies an additional symmetry relation.

Let us close this section by stating a result on h^2- expansions for symmetric linear multistep methods.

Theorem 3.66 (W.B.GRAGG [1,2], H.J.STETTER [2,3], E. HAIRER & C.LUBICH [1]): *Consider a linear multistep method which satisfies the assumptions of Theorem 3.63 resp. Theorem 3.64 and is symmetric in the above-mentioned sense. Then the asymptotic expansion of the global error is an h^2- expansion.*

Proof. Following an idea originally due to R.BULIRSCH & J.STOER [2] and further worked out by E.HAIRER, S.P.NØRSETT & G. WANNER [1], we define the *adjoint* method of (3.245) by

$$\begin{aligned} u(t_0,-h) &= \varphi(-h) \quad \text{and} \\ u(t+h,-h) &= S^{-1}\,u(t,-h) + h\,\Phi^*(t,u(t,-h),h)\,, \\ &\qquad t = t_0+\nu h, \nu \in \mathbb{Z}\,, \end{aligned}$$

where S^{-1} is assumed to exist and Φ^* is implicitly defined by the pair of formulas

$$\begin{aligned} v &= Su - h\,\Phi(t+h,u,-h)\,, \\ u &= S^{-1}v + h\,\Phi^*(t,v,h)\,. \end{aligned}$$

(This is nothing else but a clever extension of the one-step approach, cf. formulae (3.170)-(3.172).) Note that for symmetric linear multistep methods the inverse S^{-1} always exists, since here $\alpha_0 = 1$, due to (3.248a). The correct value function of the adoint method is $z(t,-h)$. Furthermore, it can be shown (cf. also the proof of Theorem 3.48) that the adjoint method possesses the same asymptotic expansion as the original one, but with h replaced by $-h$.

Now let us denote the global error by

$$e(x,h) \;:=\; u(x,h) - z(x,h)$$

and its j th component by $e_j(x,h)$.
Thanks to the relations (3.246) and (3.247) we have

$$e_k(x,-h) \;=\; e_1(x-(k-1)h,h)\,.$$

Since for any linear multistep method

$$e_k(x,h) \;=\; e_1(x-(k-1)h,h)\,,$$

we obtain immediately the relation

$$e_k(x,-h) \;=\; e_k(x,h)\;.$$

But this last component equals just $(y(x)-y_n)$, and therefore the asymptotic expansion of the global error cannot contain odd powers of h. This completes the proof of Theorem 3.66. □

The oldest and best-known example for this situation is probably the explicit midpoint rule

$$y_{\nu+2}-y_\nu \;=\; 2hf_{\nu+1}\,, \tag{3.249}$$

which is a stable symmetric linear 2-step method. We shall investigate this method in greater detail in Section 4.3.2. Here, we only remark that from Theorem 3.64 (with $k,l,q=2$) in connection with Theorem 3.66 it follows that there exist "two" asymptotic expansions, one for odd and one for even indices. The main and for a long time open problem was to find a suitable starting procedure for the calculation of y_1. It was finally proved by W.B.GRAGG [1] that taking just one Euler step as starting procedure, i.e. putting

$$\varphi(h) \;:=\; (y_0+hf_0, y_0) \tag{3.250}$$

is the right choice.

Even when applied to the *delay differential equation*

$$\begin{aligned} y'(t) &= f(t,y(t),y(t-\tau)) && \text{for } t>0\,,\\ y(t) &= \psi(t) && \text{for } -\tau\le t\le 0\,, \end{aligned}$$

the explicit midpoint rule (3.249) leads to a numerical approximation with an h^2-expansion, as was shown in a nice paper by M.DEGEE [1]. This is particularly remarkable since the true solution of this delay equation may have jump discontinuities in its derivatives in the points $t = \nu\tau$.

3.6.5 Miscellanea. In this final subsection we want to summarize some results concerning asymptotic expansions from those parts of differential theory, which we cannot treat in detail like the ones above. Please note that the only reason for our restriction to the already mentioned results is that the author of such a book is subject to several limitations of space (and time), and by far not that some results might be less important than other ones!

Let us start with some remarks concerning *second order initial value problems.* In particular, we consider problems of the form

$$\begin{aligned} y'' &= f(t,y)\ , \quad \text{with} \\ y(t_0) &= y_0\ , \quad y'(t_0) = y'_0\ , \end{aligned} \tag{3.251}$$

with prescribed real numbers y_0 and y'_0 , i.e. problems where the differential equation's right hand side does not depend on y' . For problems of this type, *Nyström methods* are known to be very efficient; let us have a closer look onto the following special one, which is known as *Störmer's (second order) rule:* With the starting procedure

$$y_1 - y_0 = h\left(y'_0 + \frac{h}{2}\,f_0\right)\ , \tag{3.252a}$$

the method reads simply

$$y_{\nu+1} - 2y_\nu + y_{\nu-1} = h^2 f_\nu \quad \nu = 1,\ldots,n-1\ . \tag{3.252b}$$

Sometimes, also a final "smoothing step" is added, cf. P.DEUFLHARD [2] or W.B.GRAGG [1,2].

Once again, it was proved the first time in W.B.GRAGG [1,2] that the final result y_n of Method (3.252) possesses a logarithmic asymptotic expansion in powers of h^2. An easy argument for this is as follows (cf. E.HAIRER, S.P.NØRSETT & G.WANNER [1]): Using the well-known technique for rewriting the second order problem (3.251) as a first order system, we obtain

$$\begin{pmatrix} y \\ y' \end{pmatrix}' = \begin{pmatrix} y' \\ f(t,y) \end{pmatrix}, \quad \text{with} \quad \begin{pmatrix} y \\ y' \end{pmatrix}(t_0) = \begin{pmatrix} y_0 \\ y_0' \end{pmatrix}. \tag{3.253}$$

Now an application of the explicit midpoint rule (3.249) with the starting procedure (3.250), using the stepsize

$$k := \frac{h}{2} = \frac{x - t_0}{2n},$$

yields the formulae

$$\begin{aligned} \eta_1 - \eta_0 &= k\,\eta_0', \\ \eta_1' - \eta_0' &= k\,f_0 \end{aligned} \tag{3.254a}$$

($\eta_0 := y_0$, $\eta_0' := y_0'$), and, for $i = 0, \ldots, 2n-2$,

$$\begin{aligned} \eta_{i+2} - \eta_i &= 2k\,\eta_{i+1}', \\ \eta_{i+2}' - \eta_i' &= 2k\,f_{i+1}. \end{aligned} \tag{3.254b}$$

This method (more precisely: the value η_{2n}) is already known to possess a logarithmic asymptotic expansion in even powers of k resp. h. (Note that, although only stated for single equations, the results on the previous pages hold of course for systems of equations as well; some minor details which have to be changed in the proofs in order to verify this are left to the reader).

Eliminating now the η_j' – terms in (3.254), we obtain the relation

$$\eta_2 - \eta_0 = 2k\,(y_0' + k\,f_0), \quad \text{and}$$

$$\eta_{i+2} - 2\eta_i + \eta_{i-2} = (2k)^2 f_i, \quad i = 0, \ldots, 2n-2,$$

from which we recognize that for the computation of η_{2n} the odd-index terms in $\{\eta_i\}$ are not needed at all. Remembering the definition of k, this proves our statement concerning Störmer's rule.

A quite challenging branch of differential theory is the theory of *stiff differential equations,* which has become recently more and more the subject of interest. Although our results from the previous subsections are not explicitly restricted to the non-stiff case, it must be said that they describe the asymptotic behaviour of the numerical methods under consideration only in a satisfactory way, if we consider non-stiff equations. The main reason for this is that, in the stiff case, the constants hidden in the " O"-terms may explode, as h tends to zero, and therefore the asymptotic expansions become worthless in practice.

In recent years, several authors have attacked the problem of establishing asymptotic expansions in the stiff case with a suitably bounded remainder term; starting probably with G. DAHLQUIST & B. LINDBERG [1] in 1973, outstanding papers and books in this context are W. AUZINGER [1,2], W. AUZINGER & R. FRANK [1], E. HAIRER & CH. LUBICH [2], E. HAIRER & G. WANNER [1] and M. VAN VELDHUIZEN [1].

As mentioned previously, in the stiff case only implicit methods have a chance to produce satisfactory results (and the other way round: their immense computational costs justifies their use only in the stiff case), and consequently the methods investigated in the context of asymptotic expansions are the implicit Euler, midpoint and trapezoidal method. As an instructive example, let us present a result from E. HAIRER & CH. LUBICH [2] (see also W. AUZINGER & R. FRANK [1]), which is typical for almost all assertions in this context.

With some real number ε, which is positive, but very close to zero, and sufficiently smooth functions f and g, we consider the (stiff) differential system

$$\begin{aligned} y' &= f(y,z), \quad y(0) = y_0, \\ \varepsilon z' &= g(y,z), \quad z(0) = z_0, \end{aligned} \tag{3.255}$$

and compute a numerical approximation in $x = t_n$ by using the implicit Euler method, which reads in this case

$$\begin{aligned} y_{\nu+1} &= y_\nu + h\, f(y_{\nu+1}, z_{\nu+1})\,, \\ \varepsilon z_{\nu+1} &= \varepsilon z_\nu + h\, g(y_{\nu+1}, z_{\nu+1})\,, \end{aligned} \tag{3.256}$$

for $\nu = 0, \ldots, n-1$, $h = x/n$. Then, under certain additional (stability) assumptions and using our notation from the previous subsection, the following result holds.

Theorem 3.67 (E. Hairer & Ch. Lubich [2]): *If the stepsize h is bounded away from zero, say $h \geq \varepsilon$, then the final result of the process (3.256) possesses the perturbed asymptotic expansion*

$$\begin{aligned} y_n &= y(x) + \sum_{\mu=1}^{m} a_\mu(x) h^\mu + O(h^{m+1}) \\ &\quad + \frac{f_z(0)}{g_z(0)} \cdot \left(I - \frac{h}{\varepsilon} g_z(0)\right)^{-1/h} V \cdot h\,\varepsilon^2 + O(h^2\varepsilon^2)\,, \\ z_n &= z(x) + \sum_{\mu=1}^{m} b_\mu(x) h^\mu + O(h^{m+1}) \\ &\quad + \left(I - \frac{h}{\varepsilon} g_z(0)\right)^{-1/h} V \cdot h\,\varepsilon + O(h^2\varepsilon)\,, \end{aligned}$$

where the functions a_μ and b_μ satisfy

$$a_\mu(0) = O(\varepsilon^2)\,, \quad b_\mu(0) = O(\varepsilon)\,, \quad \mu = 1, 2, \ldots, m\,,$$

and

$$V = \frac{g_y(0) y''(0) + g_z(0) z''(0)}{2 g_z(0)^2} + O(\varepsilon)\,.$$

The constants hidden in the "O"-terms are independent of h and ε.

For the proof of this theorem, which follows the lines of that of Theorem 3.46, as well as for more information on the precise assumptions to be made, we refer to the original paper.

Quite recently, some interesting publications concerning the numerical treatment of *differential-algebraic systems* (stiff and non-stiff case) have appeared. Here, it turns out that, very similarly to the classical theory, the standard procedures such as linear multistep methods or Runge-Kutta methods lead to approximations which possess an asymptotic expansion in powers of the step-size. The reader interested in these things is encouraged to have a look into the books by E.HAIRER, CH.LUBICH & M.ROCHE [1] and E.HAIRER & G.WANNER [1], or into the original papers such as P.DEUFLHARD, E.HAIRER & J.ZUGCK [1], CH. LUBICH [1,2] and A.OSTERMANN [1].

Unfortunately, I had to exclude the presentation of some important topics such as numerical methods for two-point boundary value problems and for partial differential equations. There are several reasons for that, the most important one being of course once again the limited space; but let us at least give references to some outstanding publications on these topics; in alphabetical order, we refer to H.BLUM [1], H.BLUM, Q.LIN & R.RANNACHER [1], K.BÖHMER [2], H.BLUM & R.RANNACHER [1,2], K.BÖHMER, P.HEMKER & H.J.STETTER [1], K.BÖHMER & J.LOCKER [1], G.FAIRWEATHER & A.V.SAYLOR [1], P.HOFMANN [1], T.LUE [1], M.MEISTER [1,2], V.PEREYRA [1], G.STARIUS [1], H.J. STETTER [1], G.WALZ [8,9] and J.WANG [1,2]. May all authors, who did not find their names on this list, although they feel that their work is at least as important as the ones just quoted, forgive me!

3.7 Problems

P 3.1. Define, for all $n \in I\!N$, $n \geq 2$, real numbers r_n and ρ_n through

$$r_n = \frac{1}{n\sin(\pi/n)} \qquad \text{and}$$

$$\rho_n = r_n \cdot \cos(\pi/n) .$$

Prove that each of the sequences $\{r_n\}$ and $\{\rho_n\}$ possesses a logarithmic asymptotic expansion with integer exponents and limit

$$\frac{1}{\pi} .$$

P 3.2. Let t denote some real parameter and z a complex variable satisfying $|z| < 2\pi$. Prove that

$$\frac{z\,\mathrm{e}^{tz}}{\mathrm{e}^z - 1} = \sum_{n=0}^{\infty} B_n(t)\,\frac{z^n}{n!} ,$$

$B_n(t)$ being the nth Bernoulli polynomial.

P 3.3. Show that all Bernoulli numbers β_n with even index do not vanish. Moreover, prove that for all $j \geq 1$

$$\operatorname{sign}(\beta_{2j}) = (-1)^{j+1} .$$

P 3.4. Consider the sequence of real numbers $\{e_n\}$, defined through

$$e_n := \sum_{\nu=1}^{n} \frac{1}{\nu} - \log(n) \qquad \text{for all } n \in I\!N .$$

The limit C of this sequence exists and is denoted as *Euler-Mascheroni constant*, its numerical value being

$$C = 0.577\,215\,664\,901\,532\,\ldots .$$

Prove that the sequence $\{e_n\}$ possesses a logarithmic asymptotic expansion of the form

$$e_n = C + \sum_{\mu=1} \frac{c_\mu}{n^\mu}$$

with

$$c_\mu = 0 \quad \text{for } \mu = 3, 5, 7, \dots .$$

Hint: Use the fact that

$$\log(n) = \int_1^n \frac{dt}{t}$$

and apply Euler's summation formula.

P 3.5. With some real parameter $\varepsilon > 0$, denote by σ_n the nth partial sum of the series

$$\sum_{\nu=1}^{\infty} \nu^{1-\varepsilon} \left(\sqrt{\nu+1} + \sqrt{\nu}\right)^{-4} .$$

Prove that σ_n possesses a logarithmic asymptotic expansion of the form

$$\sigma_n = S + \sum_{\mu=1} \frac{c_\mu}{n^{\rho_\mu}}$$

with $\rho_1 = \varepsilon$.

P 3.6. Show that Theorem 3.11 can be applied to the function

$$f(z) := \frac{\sin(\pi z)}{\pi z}$$

and derive the asymptotic expansion for the corresponding partial products.

P 3.7. Prove the following modification of Theorem 3.13: If the functions under consideration possess a Taylor expansion of degree $m+1$, then the sequence of functions $\{\sigma_n(x)\}$, defined through

$$\sigma_n(x) \;:=\; n \cdot a_1 \cdot f^{-1}\left(\frac{f(x)}{n}\right)$$

for all $n \in \mathbb{N}$, possesses a logarithmic asymptotic expansion of order m.

P 3.8. Prove that the sequence of functions $\{t_n(x)\}$, defined for all $n \in \mathbb{N}$ through

$$t_n(x) \;:=\; n \cdot \tan\left(\frac{\arctan(x)}{n}\right) ,$$

possesses a logarithmic asymptotic expansion of the form

$$t_n(x) \;=\; \arctan(x) \;+\; \sum_{\mu=1} \frac{c_\mu(x)}{n^{2\mu}} .$$

Furthermore, show that the numbers $y_i = y_i(x)$, defined for all $i \in \mathbb{N}_0$ through

$$y_0 \;:=\; x\,, \quad \text{and for } i = 1, 2, \ldots$$

$$y_i \;:=\; \frac{2 \cdot y_{i-1}}{\sqrt{1 + \sqrt{1 + (y_{i-1}/2^{i-1})^2}}}$$

satisfy

$$y_i \;=\; t_{2^i}(x) .$$

P 3.9. For the approximation of the integral

$$I(f) \;:=\; \int_0^1 f(x)dx$$

we want to use the quadrature rule

$$U_n(f) := \frac{1}{n} \cdot \sum_{\nu=1}^{2n-1} f(x_\nu)$$

where, for $n \in \mathbb{N}$,

$$x_\nu := \frac{\nu}{2n}, \quad \nu = 0, \ldots, 2n .$$

Prove that in the case of the function $f(x) := x^{-\frac{1}{2}}$, which has a singularity at the left endpoint of the interval, the following logarithmic asymptotic expansion holds (L.FOX [1]):

$$U_n(f) = I(f) + \frac{c_0}{n^{\frac{1}{2}}} + \sum_{\mu=1} \frac{c_\mu}{n^{2\mu}} .$$

Hint: Try to imitate the approach of Lemma 3.20 and Theorem 3.21.

P 3.10. Prove Corollary 3.24.
Hint: Write the integrand $h_{\alpha,\beta}(x)$ as the sum of two other functions, each of which satisfies the assumptions of Theorem 3.23. This approach is due to P.C.CHAKRAVARTI [1].

P 3.11. For some $k \in \mathbb{Z}$, let $e_k(x)$ denote the function

$$e_k(x) := \exp(2\pi i k x) .$$

Show that the application of the trapezoidal rule T_n to this function gives

$$T_n(e_k) = \begin{cases} 0, & \text{if } k/n \notin \mathbb{Z}, \\ 1, & \text{if } k/n \in \mathbb{Z}. \end{cases}$$

Hint: Use the relation

$$z^n - 1 = (z-1)(1 + z + \cdots + z^{n-1}) .$$

P 3.12. In Section 3.4, especially Theorem 3.23, we learned about handling integrands having a singularity in one of the endpoints of the integration interval by a modified trapezoidal rule, and we derived the corresponding asymptotic expansion. In addition to that, H. Rutishauser showed us – in a special case – how to modify the trapezoidal rule once again in order to derive an asymptotic expansion, which starts with the term n^{-2} (and not with $n^{-1/2}$, as would follow from Theorem 3.23).
Prove the following result from H.RUTISHAUSER [1]: For the evaluation of the singular integral

$$I := \int_0^1 x^{-\frac{1}{2}} dx$$

the numerical quadrature rule

$$R_n := \frac{1}{n}\left(t_1 f\left(\frac{t_2}{n}\right) + \sum_{\nu=1}^{n-1} f\left(\frac{\nu}{n}\right) + \frac{1}{2}\right)$$

with

$$t_1 := \sqrt{\zeta(-1/2)\cdot\zeta(1/2)} \;=\; 0.5509878 \quad \text{and}$$

$$t_2 := \frac{\zeta(-1/2)}{\zeta(1/2)} \;=\; 0.1423533$$

is used. Then the sequence $\{R_n(f)\}_{n\in\mathbb{N}}$ possesses a logarithmic asymptotic expansion of the following form:

$$R_n \;=\; I \;+\; \sum_{\mu=3} \frac{\gamma_\mu}{n^{-\frac{1}{2}+\mu}} \;+\; \sum_{\mu=1} \frac{c_\mu}{n^{2\mu}}\,.$$

Do you have an idea how to generalize this result to the case of integrands like x^α, $-1<\alpha<0$?

P 3.13. Prove the following (cf. N.E.NÖRLUND [1], C.BREZINSKI & G.WALZ [1]): Let the function g be analytic in a simply-closed region R in $\mathbb{C}$, which contains the real points $x_\nu, \ldots, x_{\nu+m}$ in its interior. Then the divided difference Δ, as defined in Definition 3.30, possesses the representation

$$\Delta(x_\nu, \ldots, x_{\nu+m}; g) = \frac{1}{2\pi i} \int_C \frac{g(z)\,dz}{(z - x_\nu) \cdots (z - x_{\nu+m})},$$

where C is some simply-closed contour in R, which contains all the points $x_\nu, \ldots, x_{\nu+m}$ in its interior.
Hint: Use representation (3.98) and the theorem of residues.

P 3.14. If a function g is analytic in a certain simply-closed region R in $\mathbb{C}$, then Cauchy's integral formula yields for all $m \in \mathbb{N}_0$

$$g^{(m)}(\xi) = \frac{m!}{2\pi i} \int_C \frac{g(z)\,dz}{(z - \xi)^{m+1}}$$

where C is some simply-closed contour in R, which contains the point ξ in its interior.
Use this and the assertion of Problem 3.13 to give – in the case of analytic functions – another proof of Lemma 3.32.

P 3.15. For real numbers $\lambda_0 < \lambda_1 < \cdots < \lambda_m$ we consider the set of functions

$$E := E_m := \{\exp(\lambda_0\, x), \ldots, \exp(\lambda_m\, x)\}$$

on a real interval $[a, b]$.
a) Prove, without using the results of Section 3.5 that E forms a complete Chebyshev system on $[a, b]$.
b) Work out explicitly the recurrence relation (3.112) for the generalized divided differences with respect to the system E, Δ_E.
c) Let ξ and η denote two distinct numbers from $[a, b]$ and g a function, belonging to the space $C^1[a, b]$. Compute the generalized divided difference $\Delta_E(\xi, \xi, \eta; g)$ and investigate its limiting behavior as η converges to ξ.

P 3.16. Prove that Relation (3.157) holds for every consistent one-step method of the form (3.151).

P 3.17. In Subsection 3.6.1 we introduced the so-called adjoint method of a one-step method, see Eqn. (3.172). What is the adjoint of Euler's method?

P 3.18. Let ϱ and σ be the characteristic polynomials of a linear multistep method and let $p \in I\!N$. Prove that the condition

$$\varrho(\exp(h)) - h\,\sigma(\exp(h)) \;=\; O(h^{p+1}) \quad \text{for} \quad h \to 0$$

as well as

$$\frac{\varrho(z)}{\log(z)} - \sigma(z) \;=\; O((z-1)^p) \quad \text{for} \quad z \to 1$$

is equivalent with the criterion (3.210a) for the method to be of order p at least.

P 3.19. Construct an explicit $k-$ step Adams method of order k.
Hint: Write the initial value problem (3.150) in integrated form

$$y(t_{\nu+k}) - y(t_{\nu+k-1}) \;=\; \int\limits_{t_{\nu+k-1}}^{t_{\nu+k}} f(\tau, y(\tau))d\tau\,,$$

and replace the integrand by the interpolating polynomial through the points (t_j, f_j), $j = \nu, \ldots, \nu+k-1$. Then apply Theorem 3.57.

P 3.20 (E.HAIRER, S.P.NØRSETT & G.WANNER [1]). Prove the following: If the functions φ, Φ and z are sufficiently differentiable, then a strictly stable general method (3.218) is of consistency order p, if and only if the conditions

$$|d_0| \;=\; O(h^p)\,, \text{ and}$$
$$|E(d_0 + \cdots + d_n) + d_{n+1}| \;=\; O(h^p)$$

for $h \to 0$ and $n \cdot h \leq$ constant are satisfied.

P 3.21. Show that for linear multistep methods cosistency in the sense of (3.211) is equivalent with "to be of cosistency order 1" in the sense of (3.225).

PART II:

LINEAR EXTRAPOLATION METHODS

4. Fundamental Concepts and General Philosophy

In the first part of this book we investigated the appearance as well as the properties of asymptotic expansions in several fields of numerical analysis; now, we are coming to the more practical aspects of our subject, that is, we will study *extrapolation methods.* Extrapolation should be viewed as a specific kind of convergence acceleration, and the connection with Part I is given by the fact that it is applicable precisely to those sequences, which possess an asymptotic expansion. In this sense, we so far established the theoretical foundations for the applicability of extrapolation.

Since it is so important for the correct understanding of the whole subject, allow me to say it once more in other words: If we want to apply extrapolation methods in order to accelerate the convergence of a given sequence, we neither need to know the exact form of the asymptotic expansion of that sequence, nor must we re-examine some theoretical considerations; instead, once we have established the existence of an asymptotic expansion (and that is precisely what we did in Part I), the pure knowledge of its type enables us to apply a suitable extrapolation process.

And *extrapolation is something very, very easy,* as we will see in a moment; therefore, it constitutes a powerful instrument for convergence acceleration.

Following this our general philosophy of keeping the extrapolation process as simple as possible, it is only natural to concentrate on *linear* extrapolation; this will be our main subject on the next pages. There exist also other, in general non-linear approaches, the most general one probably being the E-algorithm, which was found independently by C.Brezinski [2] and T.Håvie [4] (cf. on this subject also C.Brezinski & M.Redivo Zaglia [1]). We will investigate this algorithm in Section 6, and in particular we will put our linear methods into this framework. However, the major part of this book will be on the classical, i.e. linear extrapolation methods, because they are the most easy and efficient ones, see also S. Filippi [1].

In the present Section 4, we will introduce and study the elementary properties of those extrapolation processes, which fit to the different types of asymptotic expansions investigated in Part I.

4.1 Geometric Extrapolation

In this section we study processes which are designed in order to extrapolate the geometric asymptotic expansions introduced in Section 2. Instead of the more precise, but rather lengthy expression "extrapolation of sequences which possess a geometric asymptotic expansion ", we will denote this type of process by the shortened term *geometric extrapolation.*

4.1.1 The General Situation. Let us first collect some basic facts about geometric extrapolation in the general situation.

Lemma 4.1 *(Geometric extrapolation): Assume that the sequence* $\{x_n\}$ *possesses a geometric asymptotic expansion of order* m *,*

$$x_n = c_0 + \sum_{\mu=1}^{m} c_\mu \lambda_\mu^n + o(\lambda_m^n) \text{ for } n \to \infty , \qquad (4.1)$$

and choose some $K \in \mathbb{N}$, $K \leq m$. *Define for* $k = 0, \ldots, K$ *new sequences* $\{x_n^{(k)}\}_{n \in \mathbb{N}}$ *through the process*

$$x_n^{(0)} = x_n, \qquad n = 1, 2, \ldots,$$

$$x_n^{(k)} = \frac{x_{n+1}^{(k-1)} - \lambda_k x_n^{(k-1)}}{1 - \lambda_k}, \quad \begin{cases} k = 1, 2, \ldots, K, \\ n = 1, 2, \ldots . \end{cases} \tag{4.2}$$

Then each of the sequences $\{x_n^{(k)}\}_{n \in \mathbb{N}}$ *possesses a geometric asymptotic expansion of the form*

$$x_n^{(k)} = c_0 + \sum_{\mu=k+1}^{m} c_\mu^{(k)} \lambda_\mu^n + o(\lambda_m^n) \text{ for } n \to \infty \tag{4.3}$$

with certain coefficients $c_\mu^{(k)}$.

In particular, each of the sequences $\{x_n^{(k)}\}_{n \in \mathbb{N}}$ *converges faster to the limit* c_0 *than its precedessor.*

The *proof* of this lemma is very easy and left to the reader.

Recursion (4.2) constitutes the process of geometric extrapolation. As claimed above, it is a very easy technique for convergence acceleration.

Remark. Some authors prefer to write the extrapolation process in the form

$$x_n^{(k)} = x_n^{(k-1)} + \frac{x_{n+1}^{(k-1)} - x_n^{(k-1)}}{1 - \lambda_k}, \quad \begin{cases} k = 1, 2, \ldots, K, \\ n = 1, 2, \ldots . \end{cases} \tag{4.2'}$$

which is theoretically equivalent with (4.2), but sometimes differs in its numerical behavior. From (4.2′), it can be recognized quite nicely what the extrapolation process really does: the transition from one sequence $x_n^{(k-1)}$ to the next one $x_n^{(k)}$ works by adding a certain correction term (which of course tends to zero as n goes to infinity) to the first mentioned one, $x_n^{(k-1)}$.

The output of the extrapolation process is usually displayed in a triangular scheme of the following form, called a *Romberg tableau.*

$$
\begin{array}{ccccc}
x_0^{(0)} & & & & \\
 & x_0^{(1)} & & & \\
x_1^{(0)} & & \ddots & & \\
 & \vdots & & x_0^{(K-1)} & \\
\vdots & & \vdots & & x_0^{(K)} \\
 & \vdots & & x_1^{(K-1)} & \vdots \\
\vdots & & .^{.^{.}} & \vdots & \\
 & x_{K-1}^{(1)} & \vdots & & \\
x_K^{(0)} & \vdots & & & \\
\vdots & & & &
\end{array}
$$

Table 4.1. A Romberg Tableau

Let us now illustrate this extrapolation process by a first numerical example: By the end of Section 2.2 we had seen that the sequence of vectors $\{x^{[n]}\}$, defined through the application of the power method to the matrix

$$A = \begin{pmatrix} 3 & 1 & 2 \\ 1 & 4 & 5 \\ 2 & 5 & 3 \end{pmatrix}$$

possesses a geometric asymptotic expansion (cf. Formula (2.33)). This means that we can apply geometric extrapolation. The results are shown in Table 4.2: Its entries are the first components of the extrapolates (Table 4.2a) and the errors (in absolute value) of these (Table 4.2b), compared with the true number $z_1 = 0.466377389599\ldots$. The effect of the extrapolation process is evident.

1.000000				
	0.455342			
0.600000		0.462527		
	0.463835		0.466284	
0.500000		0.466019		0.466379
	0.466417		0.466384	
0.475336		0.466358		0.466377
	0.466347		0.466377	
0.468735		0.466376		0.466377
	0.466381		0.466377	
0.467006		0.466377		
	0.466377			
0.466544				

Table 4.2a. First components of the extrapolated vectors

.534 e(00)				
	.110 e(-1)			
.134 e(00)		.385 e(-2)		
	.254 e(-2)		.931 e(-4)	
.336 e(-1)		.358 e(-3)		.187 e(-5)
	.394 e(-4)		.646 e(-5)	
.896 e(-2)		.193 e(-4)		.617 e(-7)
	.299 e(-4)		.248 e(-6)	
.236 e(-2)		.159 e(-5)		.749 e(-9)
	.357 e(-5)		.128 e(-7)	
.629 e(-3)		.100 e(-6)		
	.768 e(-6)			
.166 e(-3)				

Table 4.2b. Errors of the values in Table 4.2a

There may also occur situations, where the geometric system (i.e. the set of numbers $\{\lambda_\mu\}$) is not explicitly known. Then, there is need for a suitable approximation of these numbers. The following lemma establishes one possibility for this:

Lemma 4.2: *Under the assumptions of Lemma 4.1, the sequence*

of numbers

$$\tilde{\lambda}_{n,k} := \frac{x_{n+2}^{(k-1)} - x_{n+1}^{(k-1)}}{x_{n+1}^{(k-1)} - x_n^{(k-1)}}, \tag{4.4}$$

$k \in \{1, \ldots, K\}$, *possesses a geometric asymptotic expansion of the form*

$$\tilde{\lambda}_{n,k} = \lambda_k + \sum_{\mu=1}^{\tilde{m}} \gamma_\mu \kappa_\mu^n + o(\kappa_{\tilde{m}}^n) \text{ for } n \to \infty . \tag{4.5}$$

Here, the asymptotic system $\{\kappa_\mu\} = \{\kappa_\mu(k)\}$ *as well as the order* $\tilde{m} = \tilde{m}(k)$ *depends strongly on the original asymptotic system. For example, it is*

$$\kappa_1(k) = \frac{\lambda_{k+1}}{\lambda_k} . \tag{4.6}$$

Proof. Using Relation (4.3), an elementary calculation gives us the representation

$$\tilde{\lambda}_{n,k} = \frac{\lambda_k + \sum_{\mu=k+1}^{m} \tilde{c}_\mu^{(k)} \nu_\mu^n + o(\nu_m^n)}{1 + \sum_{\mu=k+1}^{m} \hat{c}_\mu^{(k)} \nu_\mu^n + o(\nu_m^n)} \quad \text{for } n \to \infty$$

with $\nu_\mu = \lambda_\mu / \lambda_k$, and an application of Lemma 2.7 already proves the statement. □

Lemma 4.2 tells us that it is justified to replace the numbers λ_k in (4.2) by their approximations $\tilde{\lambda}_{n,k}$. What comes out is the iterative process

$$\begin{aligned} \tilde{x}_n^{(0)} &= x_n , \qquad n = 1, 2, \ldots , \\ \tilde{x}_n^{(k)} &= \frac{\tilde{x}_{n+1}^{(k-1)} - \tilde{\lambda}_{n,k} \tilde{x}_n^{(k-1)}}{1 - \tilde{\lambda}_{n,k}} , \qquad \left\{ \begin{array}{l} k = 1, 2, \ldots, K, \\ n = 1, 2, \ldots , \end{array} \right. \end{aligned} \tag{4.7}$$

sometimes denoted as *approximative geometric extrapolation.*

We apply this method to the above-mentioned example, and the resulting errors are shown in Table 4.3. Compare with Table 4.2b (and remember that for the computation of that table the exact numbers λ_μ were needed, while for Table 4.3 they are not)!

.534 e(00)				
	.289 e(-3)			
.134 e(00)		.520 e(-3)		
	.885 e(-3)		.105 e(-5)	
.336 e(-1)		.105 e(-4)		.654 e(-8)
	.553 e(-4)		.935 e(-7)	
.896 e(-2)		.123 e(-5)		.300 e(-10)
	.145 e(-4)		.534 e(-9)	
.236 e(-2)		.717 e(-7)		.763 e(-11)
	.235 e(-5)		.338 e(-10)	
.629 e(-3)		.468 e(-8)		
	.454 e(-6)			
.166 e(-3)				

Table 4.3. Errors of approximative geometric extrapolation

Inserting the definiton of $\tilde{\lambda}_{n,k}$, we see that the iteration rule of the process in (4.7) can be re-written in the form

$$\tilde{x}_n^{(k)} = \tilde{x}_n^{(k-1)} - \frac{\left(\tilde{x}_{n+1}^{(k-1)} - \tilde{x}_n^{(k-1)}\right)^2}{\tilde{x}_n^{(k-1)} - 2\,\tilde{x}_{n+1}^{(k-1)} + \tilde{x}_{n+2}^{(k-1)}} \,. \tag{4.8}$$

For $k = 1$, this is nothing else but Aitken's Δ^2-process. So, our process of approximative geometric extrapolation could be viewed as an iterative version (and thus an improvement) of the Δ^2-process.

Another very prominent generalization of this method is given by the $\epsilon-$ algorithm (or Shanks' transformation), a nonlinear sequence transformation, which also conincides in the first step with Aitken's process. The $\epsilon-$ algorithm was found by D.SHANKS [1] and put into algorithmic form by P.WYNN [2]; it is well-described in many books, e.g. P.J.DAVIS & PH.RABINOWITZ [1] or C.BREZINSKI & M.REDIVO ZAGLIA [1]. Numerical tests indicate that the approximative geometric extrapolation (4.7) resp. (4.8) is at least as efficient as the $\epsilon-$ algorithm.

For some other modifications of Aitken's process, see also P.BJØRSTADT, G.DAHLQUIST & E.GROSSE [1], R.F.KING [1], A.MEYERS [1], or P.SABLONIERRE [1].

4.1.2 Extrapolation of Special Geometric Asymptotic Expansions. Particularly in the context of iteration processes, there appear what we called *special* geometric asymptotic expansions, i.e. geometric asymptotic expansions, where all λ_μ 's are powers of one fixed number λ, $0 < \lambda < 1$, see Section 2.1. Of course, since this is a special case of the above-treated general situation, all statements from above keep true here; for example, the extrapolation process (4.2) now becomes

$$\begin{aligned} x_n^{(0)} &= x_n, \qquad n = 1, 2, \dots, \\ x_n^{(k)} &= \frac{x_{n+1}^{(k-1)} - \lambda^k x_n^{(k-1)}}{1 - \lambda^k}, \quad \begin{cases} k = 1, 2, \dots, K, \\ n = 1, 2, \dots . \end{cases} \end{aligned} \tag{4.9}$$

But, in addition, some things become easier in this special case; in particular, we have now a better way to estimate the value of λ^k resp. λ: Instead of (4.4), define

$$\tilde{\lambda}_N := \frac{x_{N+2} - x_{N+1}}{x_{N+1} - x_N} \tag{4.10}$$

for some large N, $N > m$. We had already seen that the kth power of $\tilde{\lambda}_N$ possesses a geometric asymptotic expansion with limit λ^k (cf. Problem P 2.3). Therefore, in order to define an approximative geometric extrapolation process in this special case, we can replace λ in (4.9) by $\tilde{\lambda}_N$, still keeping the asymptotic structure of the extrapolated sequences.

The advantage of using (4.10) instead of (4.4) lies clearly in the fact that the first one can be computed by means of the original sequence $\{x_n\}$ alone (in contrast to the general situation, where all $x_n^{(k)}$'s are involved).

The efficiency of the method can best be demonstrated by the following numerical example, which is taken from G. Walz [10]: Consider a real function f, defined and sufficiently smooth on the closed interval I, and having there only one zero, say ξ. Then the real function

$$T(x) := x - \frac{f(x)}{f'(x)}, \tag{4.11}$$

which is the iteration function of Newton's method, is well-defined and has ξ as a fixed point. We assume that I is small enough such that the convergence of the iterative process $x_{n+1} := T(x_n)$ is guaranteed for all $x_0 \in I$.

Let ϱ denote the multiplicity of ξ as a root of the function f. Then, as it is well-known,

$$\lambda := T'(\xi) = 1 - 1/\varrho, \tag{4.12}$$

hence $0 < T'(\xi) < 1$, whenever $\varrho \geq 2$. So we know from Theorem 2.2 that the Newton sequence $\{x_{n+1}\}$ possesses a special geometric asymptotic expansion, and the extrapolation process (4.9) can be applied. But to do this, one would have to know the exact value of λ or, in other words, of ϱ, and if this is the case, it might be better anyway to use the modified Newton method, defined by

$$T_\varrho(x) := x - \varrho \frac{f(x)}{f'(x)}, \tag{4.13}$$

which converges quadratically. However, in practice the value of ϱ will *not* be known, and so neither (4.13) nor the "exact" extrapolation process can be applied, and we have to replace λ by the approximation $\tilde{\lambda}_N$ from (4.10).

Let us apply this approach to the function

$$f(x) := (x-1)^2 \exp(x),$$

which has a double zero in $\xi = 1$, such that for the Newton iteration function

$$T(x) = \frac{x^2+1}{x+1} \tag{4.14}$$

we have $\lambda = T'(\xi) = 1/2$. But now we forget this immediately (the example is only a theoretical one anyway) and use the estimate $\tilde{\lambda}_N$ with $N = 27$ instead. The errors of the resulting extrapolation process, starting with Newton's method and $x_0 = 2$, are shown in Table 4.4. The improvement of the results compared with those of Newton's method itself (first column) is obvious.

.10 e(01)					
	.33 e(00)				
.67 e(00)		.11 e(00)			
	.17 e(00)		.30 e(-1)		
.42 e(00)		.40 e(-1)		.59 e(-2)	
	.72 e(-1)		.74 e(-2)		.75 e(-3)
.24 e(00)		.12 e(-1)		.91 e(-3)	
	.27 e(-1)		.13 e(-2)		.69 e(-4)
.14 e(00)		.26 e(-2)		.95 e(-4)	
	.86 e(-2)		.17 e(-3)		.40 e(-5)
.72 e(-1)		.47 e(-3)		.69 e(-5)	
	.25 e(-2)		.17 e(-4)		
.37 e(-1)		.74 e(-4)			
	.68 e(-3)				
.19 e(-1)					

Table 4.4. Extrapolation of a Newton sequence: Errors

Let us now turn for a moment to a more formal aspect of the extrapolation process (4.9). Although it is clear at once that the extrapolated values $x_n^{(k)}$ are linear combinations of the starting values $x_n, \ldots, x_{n+k}$, the exact determination of this combination (i.e. of its coefficients) needs some lines of calculation; this is accomplished in the following lemma.

Lemma 4.3 (G.WALZ [10]): *Let $x_n^{(k)}$ denote the output of the extrapolation process (4.9) for some k and n. Then $x_n^{(k)}$ is a linear combination of the numbers $x_n, \ldots, x_{n+k}$ of the form*

$$x_n^{(k)} = \sum_{\nu=0}^{k} \alpha_{\nu k} \cdot x_{n+\nu} \tag{4.15a}$$

where, for $\nu = 0, \ldots, k$,

$$\alpha_{\nu k} \;=\; (-1)^{k-\nu} \frac{\lambda^{k(k+1)/2}}{(1-\lambda)\cdots(1-\lambda^k)} \cdot \sum_{M(\nu,k)} \lambda^{-(j_1+\cdots+j_\nu)} \,. \quad (4.15b)$$

Here, $M(\nu, k)$ *denotes the set*

$$M(\nu,k) \;:=\; \{(j_1,\ldots,j_\nu) \in I\!N^\nu ;\, 1 \le j_1 < j_2 < \cdots < j_\nu \le k\} \,.$$

Proof. It is easy to see that $x_n^{(k)}$ is of the form

$$x_n^{(k)} \;=\; \frac{1}{(1-\lambda)\cdots(1-\lambda^k)} \cdot \sum_{\nu=0}^{k} \tilde{\alpha}_{\nu k}\, x_{n+\nu} \,,$$

where the coefficients $\tilde{\alpha}_{\nu k}$ can be computed by means of the formula

$$\begin{aligned} &\tilde{\alpha}_{0l} = (-1)^l \lambda^{l(l+1)/2} \,, \quad \tilde{\alpha}_{ll} = 1 \,, \qquad \text{and} \\ &\tilde{\alpha}_{\nu l} = \tilde{\alpha}_{\nu-1,l-1} - \lambda^l\, \tilde{\alpha}_{\nu,l-1}, \quad \nu = 1,\ldots,l-1 \end{aligned} \qquad (4.16)$$

for $l = 0, \ldots, k$. This already establishes (4.15a). Now we use induction on l (or k , resp.): Assume that the hypothesis (4.15b) is true for $l-1$ with some $l \le k$. Then, using (4.16), we obtain

$$\begin{aligned} \tilde{\alpha}_{\nu l} \;&=\; \tilde{\alpha}_{\nu-1,l-1} - \lambda^l\, \tilde{\alpha}_{\nu,l-1} \\ &=\; (-1)^{l-\nu} \lambda^{(l-1)l/2} \cdot \sum_{M(\nu-1,l-1)} \lambda^{-(j_1+\cdots+j_{\nu-1})} \\ &\qquad -\; \lambda^l \cdot (-1)^{l-1-\nu} \lambda^{(l-1)l/2} \cdot \sum_{M(\nu,l-1)} \lambda^{-(j_1+\cdots+j_\nu)} \\ &=\; (-1)^{l-\nu} \lambda^{l(l+1)/2} \cdot \\ &\qquad \cdot \left(\sum_{M(\nu-1,l-1)} \lambda^{-(j_1+\cdots+j_{\nu-1})-l} + \sum_{M(\nu,l-1)} \lambda^{-(j_1+\cdots+j_\nu)} \right) \\ &=\; (-1)^{l-\nu} \lambda^{l(l+1)/2} \cdot \sum_{M(\nu,l)} \lambda^{-(j_1+\cdots+j_\nu)} \,, \end{aligned}$$

which completes the proof of Lemma 4.3. □

Lemma 4.3 gives us more insight into the structure of the extrapolation process, and in particular it allows us to prove the following Theorem 4.4, which finally explains the name "extrapolation". Although it is not easy to trace back the statement of the theorem through the literature, it seems that F.L.BAUER, H.RUTISHAUSER & E.STIEFEL [1] as well as R.BULIRSCH [1], R.BULIRSCH & J.STOER [1,2] were the first ones who investigated such formal aspects of extrapolation processes.

Theorem 4.4: *Fix for the moment some* $k \in \mathbb{N}$ *and denote by* Λ_k *the* $(k+1)-$ *dimensional function space*

$$\Lambda_k := \operatorname{span}\{1, \lambda^t, \ldots, \lambda^{kt}\}. \tag{4.17}$$

Furthermore, let $l(x) = l(x)(t)$ *be that function from* Λ_k *, which interpolates our given sequence* $\{x_t\}_{t \in \mathbb{N}}$ *(as a function of the index) in the "points"* $t = n, n+1, \ldots, n+k$ *. Then*

$$\lim_{t \to \infty} l(x)(t) := l(x)(\infty) = x_n^{(k)}, \tag{4.18}$$

with $x_n^{(k)}$ *from (4.9).*

Remark. Theorem 4.4 states in particular that $x_n^{(k)}$ can be viewed as the extrapolation (for $t \to \infty$) of the original sequence with respect to the space Λ_k .

Proof of Theorem 4.4. By construction of the extrapolation process (4.9) in combination with Lemma 4.3 it is clear that the coefficients $\alpha_{\nu k}$ from (4.15) satisfy

$$\sum_{\nu=0}^{k} \alpha_{\nu k} \cdot \lambda^{\mu\nu} = \begin{cases} 1 & \text{for } \mu = 0, \\ 0 & \text{for } \mu = 1, \ldots, k. \end{cases} \tag{4.19}$$

Furthermore, since Λ_k is a Haar space on $[1, \infty)$, we know that there exists a uniquely determined function $l(x) \in \Lambda_k$, say

$$l(x) = \beta_0 + \beta_1 \lambda^t + \cdots + \beta_k \lambda^{kt} ,$$

which satisfies the interpolation conditions

$$\beta_0 + \beta_1 \lambda^{(n+\nu)} + \cdots + \beta_k \lambda^{k(n+\nu)} = x_{n+\nu} \tag{4.20}$$

for $\nu = 0, \ldots, k$. Now, multiply the νth line in (4.20) by $\alpha_{\nu k}$ and sum up; then, due to (4.15), the right hand side equals $x_n^{(k)}$, while the left hand side just gives β_0, thanks to the relations (4.19). Since $\lambda < 1$, the limit in (4.18) exists and equals β_0, which completes the proof of the statement. □

4.1.3 Extrapolation of Supergeometric Asymptotic Expansions. If our sequence $\{x_n\}$ is defined through an iteration process

$$x_{n+1} = T(x_n) ,$$

where now the derivative of the iteration function T in the fixed point ξ may vanish, say

$$T'(\xi) = \cdots = T^{(p-1)}(\xi) = 0 , \; T^{(p)}(\xi) \neq 0 ,$$

with $p \geq 2$, it does no longer possess a geometric asymptotic expansion. But, thanks to Theorem 2.4, we know that there exists an asymptotic expansion of the type

$$x_n = \xi + \sum_{\mu=1}^{r} c_\mu \lambda^{\mu p^n} + o(\lambda^{r p^n}) \text{ for } n \to \infty \tag{4.21}$$

with some $r = r(s,p) \in I\!N$ and $\lambda \in I\!R$, $|\lambda| < 1$, which we denoted as supergeometric asymptotic expansion.

Let us also for this case state a suited extrapolation process and an illustrating numerical example.

Lemma 4.5 (G.WALZ [10]): *Let the sequence $\{x_n\}$ possess an asympotic expansion of the form given in (4.21). Then the sequences $\{x_n^{(k)}\}$, defined by*

$$
\begin{aligned}
x_n^{(0)} &= x_n\,, \qquad n = 1,2,\dots\,,\\
x_n^{(k)} &= \frac{x_{n+1}^{(k-1)} - \lambda^{(p^k-1)p^n}\, x_n^{(k-1)}}{1-\lambda^{(p^k-1)p^n}}\,, \quad \left\{ \begin{array}{l} k = 1,2,\dots, \\ n = 1,2,\dots\,. \end{array}\right.
\end{aligned}
\tag{4.22}
$$

possess the asymptotic expansion

$$
\begin{aligned}
x_n^{(k)} &= \xi + c_{k,e(k+1)}\lambda^{e(k+1)\cdot p^n} + \cdots \\
&\qquad \cdots + c_{k,r}\lambda^{rp^n} + o(\lambda^{rp^n}) \quad \textit{for } n \to \infty\,,
\end{aligned}
\tag{4.23}
$$

where we have used the abbreviation

$$
e(\mu) := \frac{p^\mu - 1}{p-1} \qquad \textit{for } \mu \in I\!N\,.
$$

Proof. We use induction with respect to k. For $k = 0$ there is nothing to show; let now $k \geq 1$, and assume that the assertion is true for $k-1$.

Due to the induction hypothesis, $x_n^{(k)}$ can be represented in the form

$$
x_n^{(k)} = \xi + \sum_{\nu=k}^{r} \alpha_\nu(\lambda, n) + o(\lambda^{rp^n})\,, \tag{4.24}
$$

where for $\nu = k, \dots, r$

$$
\alpha_\nu(\lambda, n) = \frac{c_{k-1,\nu}}{1-\lambda^{(p^k-1)p^n}} \cdot \left(\lambda^{(e(k)+\nu-k)p^{n+1}} - \lambda^{((p^k-1)+e(k)+\nu-k)p^n}\right) \tag{4.25}
$$

with $c_{k-1,\nu}$ independent of n.

First we show that $\alpha_k(\lambda, n) = 0$; this follows from the equation

$$
\begin{aligned}
((p^k-1)+e(k))\,p^n &= ((p-1)\frac{p^k-1}{p-1} + \frac{p^k-1}{p-1})\,p^n \\
&= \frac{p^k-1}{p-1}\,p^{n+1} = e(k)\cdot p^{n+1}\,.
\end{aligned}
$$

Furthermore, since $|\lambda|$ is smaller than one, we can expand the quotient on the right hand side of (4.25) into a geometric series, and obtain

$$\alpha_\nu(\lambda, n) = \tag{4.26}$$
$$c_{k-1,\nu} \cdot (\lambda^{(e(k)+\nu-k)p^{n+1}} - \lambda^{((p^k-1)+e(k)+\nu-k)p^n}) \cdot \sum_{j=0}^{\infty} \gamma_j \lambda^{j(p^k-1)p^n}$$

with some real numbers γ_j, not depending on n. From (4.26) it follows that we can rearrange the sum in (4.24) in such a way that the desired form (4.23) is obtained.

So we are left to prove that the smallest power of λ in the new series is just $e(k+1)p^n$. To do this, we study the exponents of λ in (4.25) for $\nu = k+1$, and obtain

$$(e(k)+1)\,p^{n+1} = (\frac{p^k-1}{p-1}+1)\,p\,p^n > \frac{p^{k+1}-1}{p-1}\,p^n = e(k+1)\,p^n,$$

and, finally

$$(p^k-1+e(k)+1)\,p^n = \frac{p^k\,(p-1)+(p^k-1)}{p-1}\,p^n = e(k+1)\,p^n. \;\square$$

As a numerical example, we consider again Newton's method, this time for the computation of $\sqrt{3}$, i.e. we are concerned with the iteration function

$$T(x) = \frac{1}{2}(x + \frac{3}{x}). \tag{4.27}$$

We now have $T'(\xi) = 0$ and $T''(\xi) = 1/\sqrt{3}$, and so Lemma 4.5 (with $p = 2$) can be applied.

Again we cannot compute explicitly the value of λ^{2^n}, which is needed for the extrapolation in this case, but we can use a similar estimate for this number as in Subsection 4.1.2, since

$$\left(\frac{x_{n+2}-x_{n+1}}{x_{n+1}-x_n}\right) = \lambda^{2^n} + \tilde{c}_1\lambda^{22^n}\cdots \tag{4.28}$$

for $n \to \infty$.

In Table 4.5, we present (in Romberg form) the results of a numerical test for this case. We computed the values $x_n := T(x_{n-1})$, $n = 1, 2, \dots,$ with T from (4.27) and $x_0 = 3$, and then applied the extrapolation procedure from Lemma 4.5 to these values, with λ^{2^n} replaced by the left hand side of (4.28). The entries of the table are the errors $|\xi - x_n^{(k)}|$ for $k = 0, \dots, 3$ and $n = 0, \dots, 5 - k$.

.127 e(01)			
	.654 e(-1)		
.268 e(00)		.264 e(-3)	
	.128 e(-2)		.868 e(-8)
.179 e(-1)		.744 e(-8)	
	.474 e(-6)		.654 e(-16)
.920 e(-4)		.519 e(-17)	
	.650 e(-13)		< .1 e(-32)
.245 e(-8)		< .1 e(-32)	
	.122 e(-26)		
.173 e(-17)			

Table 4.5. Extrapolation of a supergeometric asymptotic expansion: Errors

4.2 Logarithmic Extrapolation

We will now consider those extrapolation processes, which are connected with the logarithmic asymptotic expansions treated in Section 3; as before, we shall abbreviate the term "extrapolation of logarithmic asymptotic expansions" and just say *logarithmic extrapolation* instead. In analogy to the previous section, we will first present the procedure of the extrapolation process and some numerical examples illustrating it, and then supply an interesting connection with interpolation theory, which explains in particular why the process (4.31) is bearing the name extrapolation.

4.2.1 Extrapolation of Logarithmic Asymptotic Expansions: The Procedure. Consider a sequence of numbers $\{\sigma_n\}$, which possesses a logarithmic asymptotic expansion of order m. We recall from Part I that each element σ_n can therefore be written in the form

$$\sigma_n = c_0 + \sum_{\mu=1}^{m} \frac{c_\mu}{n^{\rho_\mu}} + o(n^{-Re\rho_m}) \quad \text{for } n \to \infty \tag{4.29}$$

with coefficients $\{c_\mu\}$ independent of n (see Definition 3.1 for details).

The basic idea of extrapolation is now once again to compute the values of σ_n for several choices of n, say $n = n_0 < n_1 < n_2 < \cdots$, and to combine them in order to obtain new sequences, which converge faster than the original ones. For most applications it is convenient to choose the sequence $\{n_i\}$ not just anyhow, but as a geometric progression (there might, however, occur situations where this is not general enough; then we will have to make additional asumptions on the exponents ρ_μ, see Subsection 4.2.3): Choose natural numbers n_0 and b, $b \geq 2$, and put

$$n_i := n_0 \cdot b^i, \qquad i = 0, 1, 2, \dots . \tag{4.30}$$

Then the logarithmic asymptotic expansion in (4.29) has changed into a special geometric one, and the extrapolation process reads as follows:

Lemma 4.6 *(Logarithmic extrapolation): Let there be given a sequence $\{\sigma_n\}$, which possesses a logarithmic asymptotic expansion of the from (4.29), and a sequence of natural numbers $\{n_i\}$, satisfying (4.30). Furthermore, choose some $K \in \mathbb{N}$, $K \leq m$ and define for $k = 0, \dots, K$ new sequences $\{y_i^{(k)}\}_{i \in \mathbb{N}}$ through the process*

$$\begin{aligned} y_i^{(0)} &= \sigma_{n_i}, \qquad i = 0, 1, \dots, \\ y_i^{(k)} &= \frac{b^{\rho_k} \cdot y_{i+1}^{(k-1)} - y_i^{(k-1)}}{b^{\rho_k} - 1}, \quad \begin{cases} k = 1, 2, \dots, K, \\ i = 0, 1, \dots , \end{cases} \end{aligned} \tag{4.31}$$

or, equivalently,

$$y_i^{(k)} = y_{i+1}^{(k-1)} + \frac{y_{i+1}^{(k-1)} - y_i^{(k-1)}}{b^{\rho_k} - 1}, \quad \begin{cases} k = 1, 2, \ldots, K, \\ i = 0, 1, \ldots \, . \end{cases} \tag{4.31'}$$

Then each of the sequences $\{y_i^{(k)}\}_{i \in \mathbb{N}}$ *possesses a logarithmic asymptotic expansion of the form*

$$y_i^{(k)} = c_0 + \sum_{\mu=k+1}^{m} \frac{c_\mu^{(k)}}{n_i^{\rho_\mu}} + o(n_i^{-Re\rho_m}) \quad \text{for } n_i \to \infty \tag{4.32a}$$

with coefficients

$$c_\mu^{(k)} = c_\mu \cdot \prod_{j=1}^{k} \frac{b^{\rho_j - \rho_\mu} - 1}{b^{\rho_j} - 1} \quad \text{for } k \geq 1 \text{ and } \mu \geq k+1, \tag{4.32b}$$

which are independent of n_i *.*

In particular, each of the sequences $\{y_i^{(k)}\}$ *converges faster to the limit* c_0 *than its precedessor.*

Recursion (4.31) constitutes the process of logarithmic extrapolation. Once again, we leave the very easy *proof* of this lemma to the reader (note that $n_{i+1} = b\, n_i$, and use induction on k).

As in the case of geometric extrapolation, it is very difficult to decide who was the first one publishing the statement of Lemma 4.6. In a certain sense, the story indeed begins with ARCHIMEDES [1]; other names, which are strongly connected with (sometimes special cases of) the extrapolation process (4.31) are R. Bulirsch, Ch. Huygens, K. Kommerell, H. Rutishauser, J.F. Saigey, J. Stoer, C. Schneider, and, of course, L.F. Richardson and W. Romberg. For an extension of this list and more detailed information on it see our Historical Notes or, for example, the publications by C.BREZINSKI [6], J.DUTKA [1], H.ENGELS [2], D.C.JOYCE [1] and G.WALZ [1,11].

In most applications, one is of course interested in computing as few values of σ_n and with the smallest numbers n_i as possible. Therefore, putting

$$b := 2 \quad \text{and} \quad n_0 := 1 \tag{4.33}$$

in (4.30) is the most popular choice.

Within the context of numerical quadrature of non-smooth integrands like $x^\beta \log(x)$, we had seen in Part I (cf. Theorem 3.21) that the asymptotic expansion of the trapezoidal rule might contain a term of the form

$$\frac{\gamma}{n^{1+\beta} \log(1/n)^{-1}} . \tag{4.34}$$

This means that this asymptotic expansion is no longer a logarithmic one, which implies that the corresponding extrapolation process must be slightly modified; for example, following L.Fox [1], one can choose $n_i = 2^i$ and use the procedure

$$y_i^{(1)} = \frac{i\,2^{1+\beta}\,y_{i+1}^{(0)} - (i+1)\,y_i^{(0)}}{i\,(2^{1+\beta}-1)-1} ,$$

in order to eliminate terms of the form given in (4.34).

4.2.2 Numerical Examples. Since the process of logarithmic extrapolation as introduced in Lemma 4.6 is so fundamental and important, let us take some time and space to study several numerical illustrations for it, at the same time continuing some examples from Part I.

Example 1. We begin with the sequence

$$\sigma_n^e = \sigma_n^e(z) = \left(1 + \frac{z}{n}\right)^n ,$$

z any fixed complex number, which we had proved in Lemma 3.45 to possess a logarithmic asymptotic expansion of arbitrary order with $\rho_\mu = \mu$ for all μ. Hence the logarithmic extrapolation process (4.31) can be applied; in this case, it takes the simple form

$$
\begin{aligned}
&y_i^{(0)}(z) = \sigma_{2^i}^e(z)\,, \qquad\qquad i = 0, 1, \dots\,, \\
&y_i^{(k)}(z) = \frac{2^k \cdot y_{i+1}^{(k-1)}(z) - y_i^{(k-1)}(z)}{2^k - 1}\,, \quad \begin{cases} k = 1, 2, \dots, \\ i = 0, 1, \dots\,, \end{cases}
\end{aligned}
\tag{4.35}
$$

where we have regarded (4.33). In the usual form of a Romberg tableau (cf. Table 4.1), Table 4.6 shows the *errors* (in absolute value) of the approximations $y_i^{(k)}(1)$ for $k = 0, \dots, 7$ and $i = 0, \dots, 7-k$.

Note that the first column, which contains the errors of the original sequence $\{\sigma_n^e(1)\}$ for $n = 2^0, 2^1, \dots, 2^7$, ends up with a value still greater than 10^{-2}, while the rightmost entry of the table, which comes from nothing else but an application of the simple extrapolation process (4.35) to this first column, is already below 10^{-9}! This explains a corresponding remark in the introduction.

.72 e(00)							
	.22 e(00)						
.47 e(00)		.41 e(-1)					
	.85 e(-1)		.44 e(-2)				
.28 e(00)		.90 e(-2)		.25 e(-3)			
	.28 e(-1)		.51 e(-3)		.75 e(-5)		
.15 e(00)		.16 e(-2)		.15 e(-4)		.11 e(-6)	
	.82 e(-2)		.46 e(-4)		.23 e(-6)		.87 e(-9)
.80 e(-1)		.24 e(-3)		.69 e(-6)		.17 e(-8)	
	.22 e(-2)		.35 e(-5)		.53 e(-8)		
.41 e(-1)		.33 e(-4)		.27 e(-7)			
	.58 e(-3)		.25 e(-6)				
.21 e(-1)		.43 e(-5)					
	.15 e(-3)						
.11 e(-1)							

Table 4.6. Errors $|y_i^{(k)}(1) - \mathrm{e}^1|$, where $y_i^{(0)}(1) = (1 + 2^{-i})^{2^i}$

Example 2. With our second example we again take up another one from Subsection 3.4.2, which had served as an illustration of Theorem 3.21. There, we had seen as a special case of a more general result that the trapezoidal rule with n subintervals, applied to the function $x^{0.25}$ on the interval $[0,2]$, produces a sequence of approximations, say $T_n(x^{0.25})$, with the logarithmic asymptotic expansion

$$T_n(x^{0.25}) \;=\; \int_0^2 x^{0.25}dx \;+\; \frac{\gamma_1}{n^{1.25}} \;+\; \sum_{\mu=1} \frac{c_\mu}{n^{2\mu}} \tag{4.36}$$

of arbitrary order.

Hence, we can apply the extrapolation process (4.31) (with $\rho_1 = 1.25$ and $\rho_\mu = 2\mu - 2$, $\mu = 2, 3, \ldots$) to this sequence, thus obtaining faster converging new ones $y_i^{(k)}$. This is illustrated in Table 4.7; the entries of Part a are once again the errors of these extrapolated sequences. Compare the errors in the last column ($k = 5$) with that of the trapezoidal rule itself (first column)!

.714 e(00)					
	.140 e(-1)				
.308 e(00)		.130 e(-3)			
	.360 e(-2)		.163 e(-5)		
.132 e(00)		.968 e(-5)		.121 e(-7)	
	.908 e(-3)		.374 e(-7)		.410 e(-10)
.559 e(-1)		.640 e(-6)		.881 e(-10)	
	.228 e(-3)		.671 e(-9)		.955 e(-13)
.236 e(-1)		.406 e(-7)		.439 e(-12)	
	.569 e(-4)		.109 e(-10)		.134 e(-15)
.997 e(-2)		.255 e(-8)		.185 e(-14)	
	.142 e(-4)		.173 e(-12)		.148 e(-18)
.420 e(-2)		.160 e(-9)		.737 e(-17)	
	.356 e(-5)		.270 e(-14)		
.177 e(-2)		.997 e(-11)			
	.889 e(-6)				
.744 e(-3)					

Table 4.7a. Extrapolation of the Trapezoidal Rule for $x^{0.25}$ (Errors)

In some sense even more interesting is Table 4.7b; here, the entries of the kth column are the quotients of two subsequent errors in that column, i.e. the numbers

$$\frac{e_i^{(k)}}{e_{i+1}^{(k)}}\,, \qquad i = 0, 1, \dots\,,$$

where, for all i,

$$e_i^{(k)} \;=\; y_i^{(k)} - \int_0^2 x^{0.25}\,dx\;.$$

As expected (cf. also Problem P4.5), the values in the kth column converge to 2^{ρ_k+1}.

.232 e(01)					
	.389 e(01)				
.234 e(01)		.135 e(02)			
	.397 e(01)		.436 e(02)		
.236 e(01)		.151 e(02)		.137 e(03)	
	.399 e(01)		.557 e(02)		.429 e(03)
.237 e(01)		.158 e(02)		.201 e(03)	
	.400 e(01)		.615 e(02)		.711 e(03)
.237 e(01)		.159 e(02)		.237 e(03)	
	.400 e(01)		.633 e(02)		.910 e(03)
.237 e(01)		.160 e(02)		.251 e(03)	
	.400 e(01)		.638 e(02)		
.238 e(01)		.160 e(02)			
	.400 e(01)				
.238 e(01)					

Table 4.7b. Quotients of the values in Table 4.7a

Example 3. In this our last example we consider the numerical evaluation of Riemann's Zeta-Function; it was proved in Theorem 3.8 that for all $s \in \mathbb{C}$ with $\mathrm{Re}(s) > 1$ the sequence of partial sums

$$\zeta_n(s) \;=\; \sum_{\nu=1}^{n} \frac{1}{\nu^s}$$

possesses the logarithmic asymptotic expansion

$$\zeta_n(s) = \zeta(s) - \frac{(s-1)^{-1}}{n^{s-1}} + \frac{2^{-1}}{n^s} + \sum_{\mu=1} \frac{c_{2\mu}(s)}{n^{s+2\mu-1}} \tag{4.37}$$

of arbitrary order, where the $c_{2\mu}(s)$ are explicitly known. We note in passing that this is one of the rare cases known so far, where the sequence of exponents ρ_μ is indeed non-real.

For instructive reasons we computed the numbers $\zeta_n(2)$ for $n = 1, 2, 4, 8, \ldots, 1024$ and applied the logarithmic extrapolation process (4.31). Results are presented in Table 4.8; this time, to give the reader an expression of "what happens", we show the computed approximations $y_i^{(k)}$ for $k = 0, 3, 6$ themselves (to 15 decimals). Compare these with the true value

$$\zeta(2) = \frac{\pi^2}{6} = 1.644934066848226\ldots$$

and compute, as an exercise, the sequence of error-quotients as in the previous example!

$k = 0$	$k = 3$	$k = 6$
1.000000000000000	1.644418529316489	1.644934066663747
1.250000000000000	1.644909465495229	1.644934066847944
1.423611111111111	1.644933169862643	1.644934066848226
1.527422052154195	1.644934037518127	1.644934066848226
1.584346533444987	1.644934065920658	1.644934066848226
1.614167262827924	1.644934066819152	
1.629430501408888	1.644934066847317	
1.637152004954462	1.644934066848198	
1.641035436308681		
1.642982847955097		
1.643957981030164		

Table 4.8. Approximations for $\zeta(2)$

4.2.3 The Name "Extrapolation"; Generalization to Other Index Sequences. In a way very similar to that in Subsection 4.1.2 for special geometric extrapolation it is also possible to prove the following result, which finally allows us to explain why the process (4.31) is bearing the name "extrapolation".

Lemma 4.7 (cf. R.BULIRSCH & J.STOER [1], J.C.MILLER [1], G.WALZ [1]): *With $k \in \mathbb{N}$, let $\tilde{\Pi}_k$ denote the $(k+1)$-dimensional space of generalized polynomials*

$$\tilde{\Pi}_k := \operatorname{span}\{1, t^{-\rho_1}, \ldots, t^{-\rho_k}\}. \tag{4.38}$$

Furthermore, let $\pi(\sigma) = \pi(\sigma)(t)$ be that function from $\tilde{\Pi}_k$, which interpolates our given sequence $\{\sigma_t\}_{t \in \mathbb{N}}$ (as a function of the index) in the "points" $t = b^i, b^{i+1}, \ldots, b^{i+k}$. Then

$$\lim_{t \to \infty} \pi(\sigma)(t) := \pi(\sigma)(\infty) = y_i^{(k)}, \tag{4.39}$$

with $y_i^{(k)}$ from (4.31).

Sketch of proof. It is clear that $y_i^{(k)}$ is a linear combination of the σ_n's of the form

$$y_i^{(k)} = \sum_{\nu=0}^{k} \gamma_{\nu k} \cdot \sigma_{n_{i+\nu}}, \tag{4.40}$$

with certain coefficients $\gamma_{\nu k}$, which satisfy by construction the relation

$$\sum_{\nu=0}^{k} \gamma_{\nu k} \cdot b^{-(i+\nu)\rho_\mu} = \begin{cases} 1 & \text{for } \mu = 0, \\ 0 & \text{for } \mu = 1, \ldots, k \end{cases} \tag{4.41}$$

where we put $\rho_0 := 0$.

From this point on the proof runs the same lines as that of Theorem 4.4 and is therefore omitted here. □

We learn from Property (4.41) that when we deal with the extrapolation process (4.31), we have it in fact to do with a certain kind of interpolation problem, and the statement of Lemma 4.7 means that $y_i^{(k)}$ is just the value of this interpolation problem's solution in $t = \infty$. This is the reason for denoting the process (4.31) as "extrapolation ".

In some (rare) cases it might happen that the sequence $\{\sigma_n\}$ under consideration is known to possess a logarithmic asymptotic expansion, but the ρ_μ 's are not known explicitly; in E.Christiansen & H.G.Petersen [1], an algorithm is proposed, which copes with such situations.

Now it is natural to ask for the existence of a generalization of our extrapolation method for the case of more general index sequences, i.e. for n_i 's which do not necessarily satisfy (4.30). If we view it for a moment as a problem of interpolation only (i.e. without asking for the existence of an asymptotic expansion for the extrapolated sequences), this question can partially be answered with "yes".

"Partially" means in this context that we have to impose the additional assumption that all exponents ρ_μ are of the form

$$\rho_\mu = \eta \cdot \mu \tag{4.42}$$

with some fixed real number $\eta > 0$; in Part I (cf. Definition 3.40) we had such expansions denoted as *special* logarithmic asymptotic expansions.

The most general case, i.e. if both the n_i 's and the ρ_μ 's are arbitrarily spaced, can not be treated in such an elementary way as follows, but there is a possibility to consider it as special case of Brezinski's E-algorithm (see Section 6), which is, however, non-linear in general.

Lemma 4.8 (R.Bulirsch & J.Stoer [1]): *Let there be given a sequence of natural numbers*

$$n_0 < n_1 < \cdots < n_i < \cdots$$

and a set of prescribed real values $\{z_0, z_1, \ldots, z_i, \ldots\}$ *. Furthermore, consider for fixed* $\eta > 0$ *and* $k \in I\!N$ *the* $(k+1)-$ *dimensional space of generalized polynomials*

$$\hat{\Pi}_k \;:=\; \operatorname{span}\{1, t^{-\eta}, t^{-2\eta}, \ldots, t^{-k\eta}\} \;\subset\; C[1,\infty)\,, \tag{4.43}$$

and denote by $\hat{y}_i^{(k)}(t)$ *the uniquely determined function from this space, which satisfies the interpolation conditions*

$$\hat{y}_i^{(k)}(n_\mu) \;=\; z_\mu \quad \textit{for} \;\; \mu = i, \ldots, i+k\,. \tag{4.44}$$

Then the value of $\hat{y}_i^{(k)}(t)$ *at* $t = \infty$ *, say* $Y_i^{(k)}$ *, can be computed by means of the recursion*

$$\begin{aligned} Y_i^{(0)} &= z_i\,, \qquad i = 0,1,\ldots\,, \\ Y_i^{(k)} &= \frac{n_{i+k}^\eta Y_{i+1}^{(k-1)} - n_i^\eta Y_i^{(k-1)}}{n_{i+k}^\eta - n_i^\eta}\,, \quad \left\{ \begin{array}{l} k = 1,2,\ldots, \\ i = 0,1,\ldots, \end{array} \right. \end{aligned} \tag{4.45}$$

or, equivalently,

$$Y_i^{(k)} \;=\; Y_{i+1}^{(k-1)} + \frac{Y_{i+1}^{(k-1)} - Y_i^{(k-1)}}{\left(\frac{n_{i+k}}{n_i}\right)^\eta - 1}\,, \quad \left\{ \begin{array}{l} k = 1,2,\ldots, \\ i = 0,1,\ldots\,. \end{array} \right. \tag{4.45'}$$

Proof. Putting $h_\mu := n_\mu^{-1}$, the following problem is equivalent to the one stated in the lemma: Determine the value $H_i^{(k)}(0)$ of the polynomial

$$H_i^{(k)}(t) \;\in\; \operatorname{span}\{1, t^\eta, t^{2\eta}, \ldots, t^{k\eta}\} \;\subset\; C(0,1]\,,$$

which is uniquely determined through the interpolation conditions

$$H_i^{(k)}(h_\mu) \;=\; z_\mu \quad \text{for} \;\; \mu = i, \ldots, i+k\,.$$

But this newly found problem can be solved easily by an application of the Neville-Aitken interpolation process, to be found in any textbook on Numerical Analysis, e.g. J.STOER & R.BULIRSCH [1]:

$$H_i^{(0)}(0) = z_i\,, \qquad i = 0, 1, \ldots\,, \tag{4.46}$$

$$H_i^{(k)}(0) = \frac{h_i^\eta \cdot H_{i+1}^{(k-1)}(0) - h_{i+k}^\eta H_i^{(k-1)}(0)}{h_i^\eta - h_{i+k}^\eta}\,, \quad \begin{cases} k = 1, 2, \ldots, \\ i = 0, 1, \ldots\,. \end{cases}$$

The equivalence of (4.45) with (4.46) finally proves Lemma 4.8. □

Remark. The statement of Lemma 4.8 obviously remains true if we take an arbitrary set of mutually prime numbers $\{n_i\}$ (i.e. not necessarily an ascending one). In W.BARON [1], some results on the behaviour of the extrapolation process in the case of such an arbitrary index sequence are presented.

Corollary 4.9 (R.BULIRSCH & J.STOER [1]): *If the z_i's in Lemma 4.8 come from a special logarithmic asymptotic expansion, i.e. if there is a sequence $\{\sigma_n\}$ with*

$$\sigma_n = c_0 + \sum_{\mu=1}^{m} \frac{c_\mu}{n^{\eta\cdot\mu}} + o(n^{-\eta\cdot m}) \quad \text{for } n \to \infty\,, \tag{4.47}$$

such that $z_i = \sigma_{n_i}$ for all i, then the process (4.45) eliminates the terms for $\mu = 1, 2, \ldots, k$ in (4.47). Therefore, (4.45) can be viewed as a generalization resp. modification of the logarithmic extrapolation process (4.31).

Proof. For $1 \le \mu \le k$, all terms of the form $\frac{c_\mu}{n^{\eta\cdot\mu}}$ are elements of the space $\hat{\Pi}_k$ in the variable n; therefore, the statement follows from Lemma 4.8. □

Remark. For the special case $\eta = 2$ and $k = 1$, Process (4.45) was introduced (in terms of $h = n^{-1}$) by L.F.RICHARDSON [1] in 1910 and denoted by himself as h^2- extrapolation. This is why many authors speak of *Richardson-extrapolation* in this context.

Although the extrapolation process (4.45) is in this form not applicable to asymptotic expansions with general exponents ρ_μ (cf. e.g. our examples 2 and 3 in the previous subsection), it has become very popular for specific applications, mostly within the theory of ordinary differential equations, where the logarithmic asymptotic expansions are almost exclusively of the special type (4.47), see Section 3.6.

Besides the so-called *Romberg sequence*

$$\{1,2,4,8,16,\ldots\}\,, \quad \text{i.e. } n_i = 2^i\,,$$

for which we recover (4.31) with (4.33), the most frequently used index sequences in (4.45) are the *Bulirsch sequence*

$$\{1,2,3,4,6,8,12,\ldots\}\,, \quad \text{i.e.}$$

$$n_0 = 1 \text{ and } n_i = \begin{cases} 3\cdot 2^{\frac{i-2}{2}} & \text{if } i \text{ is even,} \\ 2^{\frac{i+1}{2}} & \text{if } i \text{ is odd,} \end{cases}$$

(R.BULIRSCH [1]) and, to some extend, the *Harmonic sequence*

$$\{1,2,3,4,\ldots\}\,, \quad \text{i.e. } n_i = i+1\,.$$

Moreover, some algorithms, in particular the ODE-solvers, use variable stepsizes, i.e. the choice of the stepsizes $\{h_i\}$ (resp. the indices $\{n_i\}$) is governed by some *stepsize control* mechanism; for detailed information on this topic, see for example P.DEUFLHARD [1] or L.F.SHAMPINE [1,2,3].

In view of the fact that the generalized extrapolation process (4.45) was established as the solution of an interpolation problem for (generalized) polynomials, it seems only natural to ask for the solution of the same problem when using other interpolation spaces, e.g. rational functions. Building on some earlier work of J.STOER [1], this was attacked and solved in

R.BULIRSCH & J.STOER [1,5] for arbitrary combinations of the numerator's and denominator's degree. The formulas for this general case are rather complicated and rarely used in practice; but they simplify considerably, if we make the following special choice: Denote the numerator's and denominator's degree by $\delta(n)$ and $\delta(d)$ respectively, and put, for given $k \in \mathbb{N}_0$,

$$(\delta(n), \delta(d)) := \begin{cases} (\frac{k}{2}, \frac{k}{2}), & \text{if } k \text{ is even, and} \\ (\frac{k-1}{2}, \frac{k+1}{2}), & \text{if } k \text{ is odd.} \end{cases} \tag{4.48}$$

Then consider the problem stated in Lemma 4.8, but with the (generalized) polynomial space $\hat{\Pi}_k$ replaced by the space of (generalized) rational functions with the degrees chosen subject to (4.48). Then (4.45') changes to

$$\begin{aligned} Y_i^{(-1)} &= 0, \qquad i = 0, 1, \dots, \\ Y_i^{(0)} &= z_i, \qquad i = 0, 1, \dots, \\ Y_i^{(k)} &= Y_{i+1}^{(k-1)} + \frac{Y_{i+1}^{(k-1)} - Y_i^{(k-1)}}{\left(\frac{n_{i+k}}{n_i}\right)^{\eta} \cdot \left(1 - \frac{Y_{i+1}^{(k-1)} - Y_i^{(k-1)}}{Y_{i+1}^{(k-1)} - Y_{i+1}^{(k-2)}}\right) - 1}, \quad \begin{cases} k = 1, 2, \dots, \\ i = 0, 1, \dots . \end{cases} \end{aligned} \tag{4.49}$$

Another approach to rational extrapolation using continued fractions is presented in L.WUYTACK [1]. In the 60's and 70's, these rational extrapolation methods became very popular, but later it turned out that it might not be preferable to polynomial (i.e. linear) extrapolation, regarding computational costs and stability, see e.g. S.FILIPPI [1], P.DEUFLHARD [1], or E.HAIRER, S.P.NØRSETT & G.WANNER [1].

When we ask for the convergence acceleration properties of (4.45), we should remember that this process established the solution of a certain *interpolation* (and not a convergence acceleration) problem. In other words, we have lost the nice property of the extrapolation process (4.31) that all sequences $\{y_i^{(k)}\}$ possess an asymptotic expansion. Instead, the following result holds:

Theorem 4.10 (F.L.BAUER, H.RUTISHAUSER & E.STIEFEL [1], R.BULIRSCH [1], R.BULIRSCH & J.STOER [1], P.-J.LAURENT [1,2]): *Consider a sequence $\{\sigma_n\}$ possessing a special logarithmic* *asymptotic expansion of order* m *, i.e.*

$$\sigma_n = c_0 + \sum_{\mu=1}^{m} \frac{c_\mu}{n^{\eta\cdot\mu}} + o(n^{-\eta\cdot m}) \quad \text{for } n \to \infty .$$

Furthermore, fix some constant α *,* $0 < \alpha < 1$ *, and choose a sequence of indices* $\{n_i\}$ *, such that for all* i

$$\frac{n_i}{n_{i+1}} \leq \alpha . \tag{4.50}$$

Then for each fixed $k < m$ *the asymptotic estimate*

$$Y_i^{(k)} = c_0 + O((n_i \cdots n_{i+k})^{-\eta}) \quad \text{for } i \to \infty \tag{4.51}$$

holds.

Remark. Relation (4.50) is satisfied by the most important sequences such as Romberg's or Bulirsch's; it is, however, not satisfied by the Harmonic sequence, where in general weaker results hold, see e.g. R.BULIRSCH & J.STOER [1].

Proof of Theorem 4.10. Let us use the abbreviations $h := n^{-\eta}$ and $h_\nu := n_\nu^{-\eta}$ for all ν. Then the space $\hat{\Pi}_k$ of generalized polynomials in the variable n, as defined in (4.43), coincides with the space of algebraic polynomials Π_k in h. Furthermore, without any restriction of generality, let us treat the case $i = 0$ only (we shall have indices enough to struggle with).

We first conclude from Lemma 4.8 that $Y_0^{(k)}$ is the value in $h = 0$ of that polynomial, which interpolates the given values $z_0, \ldots, z_k$ $(= \sigma_{n_0}, \ldots, \sigma_{n_k})$. Consequently, $Y_0^{(k)}$ can be written as

$$Y_0^{(k)} = \sum_{\nu=0}^{k} L_\nu^{(k)} \cdot \sigma_{n_\nu} \tag{4.52a}$$

with the Lagrange numbers

$$L_\nu^{(k)} = \prod_{\substack{\mu=0\\ \mu\neq\nu}}^{k} \frac{h_\mu}{h_\mu - h_\nu} . \tag{4.52b}$$

Now we make use of Corollary 4.9, which implies that

$$\left.\begin{aligned} Y_0^{(k)} &= \sum_{\nu=0}^{k} L_\nu^{(k)} \Big(c_0 + \sum_{\mu=1}^{m} c_\mu h_\nu^\mu + o(h^m) \Big) \\ &= \sum_{\nu=0}^{k} L_\nu^{(k)} c_0 + \sum_{\mu=1}^{m} c_\mu \sum_{\nu=0}^{k} L_\nu^{(k)} h_\nu^\mu + o(h^m) \\ &= c_0 + c_{k+1} \sum_{\nu=0}^{k} L_\nu^{(k)} h_\nu^{k+1} + o(h^{k+1}) \end{aligned}\right\} \tag{4.53}$$

We now claim that there is exists a constant C, depending on α, η, and k, but not on $\{n_i\}$, such that

$$\sum_{\nu=0}^{k} |L_\nu^{(k)}| h_\nu^{k+1} \leq C \cdot h_0 \cdots h_k . \tag{4.54}$$

Obviously, in connection with (4.53) this will prove the statement of Theorem 4.10.

Now we establish the following chain of equalities

$$\sum_{\nu=0}^{k} |L_\nu^{(k)}|\, h_\nu^{k+1} = \sum_{\nu=0}^{k} \prod_{\substack{\mu=0\\ \mu\neq\nu}}^{k} \frac{h_\mu}{|h_\mu - h_\nu|} \cdot h_\nu^{k+1} =$$

$$= h_0 \cdots h_k \cdot \sum_{\nu=0}^{k} h_\nu^k \cdot \prod_{\substack{\mu=0 \\ \mu \neq \nu}}^{k} \frac{1}{|h_\mu - h_\nu|}$$

$$= h_0 \cdots h_k \cdot \sum_{\nu=0}^{k} \prod_{\mu=0}^{\nu-1} \frac{h_\nu}{h_\mu - h_\nu} \cdot \prod_{\mu=\nu+1}^{k} \frac{h_\nu}{h_\nu - h_\mu}$$

$$= h_0 \cdots h_k \cdot \sum_{\nu=0}^{k} \prod_{\mu=0}^{\nu-1} \frac{1}{h_\mu / h_\nu - 1} \cdot \prod_{\mu=\nu+1}^{k} \frac{1}{1 - h_\mu / h_\nu}$$

$$\leq h_0 \cdots h_k \cdot \sum_{\nu=0}^{k} \prod_{\mu=0}^{\nu-1} \frac{1}{\alpha^{(\nu-\mu)\eta} - 1} \cdot \prod_{\mu=\nu+1}^{k} \frac{1}{1 - \alpha^{(\mu-\nu)\eta}},$$

which finally verifies Relation (4.54). □

4.3 Some Specific Numerical Methods as Special Cases of Logarithmic Extrapolation

It is in some sense quite natural that there exist some particular methods and algorithms, which are well-known to each mathematician, but what is not so much known is the fact that they can be considered as special cases of the extrapolation processes just introduced. In this section we want to investigate shortly a few outstanding numerical methods, which turn out to be special cases in the sense just mentioned.

4.3.1 Romberg Quadrature. In 1955, W. Romberg published a short paper on "simplified numerical integration" (W. ROMBERG [1]). In an a little bit modernized notation, the fundamental idea of that paper can be stated as follows:

Consider the numerical integration of some function $f \in C^{2r+1}[a,b]$, $r \in \mathbb{N}_0$. To do this, define, for natural numbers

$n_i = 2^i$, the trapezoidal values

$$T_i^0 := T_{n_i}(f) = \frac{b-a}{n_i}\left(\frac{1}{2}\,f(a) + \sum_{\nu=1}^{n_i-1} f\left(a+\nu\,\frac{b-a}{n_i}\right) + \frac{1}{2}\,f(b)\right) \tag{4.55}$$

(cf. (3.54)), and the midpoint values

$$U_i^0 := U_{n_i}(f) = \frac{b-a}{n_i}\cdot\sum_{\nu=0}^{n_i-1} f\left(a+(2\nu+1)\,\frac{b-a}{2n_i}\right), \tag{4.56}$$

both of which's error is of order n_i^{-2} .

Romberg first observed that for each $n \in I\!N$ the relation

$$T_{2n}(f) \;=\; \frac{T_n(f)+U_n(f)}{2} \tag{4.57}$$

holds, which obviously reduces the amount of work for the computation of the sequences $\{T_n(f)\}$ and $\{U_n(f)\}$ considerably.

Then he stated the result (refering to the monograph by L. Collatz [1]) that, if we define new sequences $\{T_i^k\}$ and $\{U_i^k\}$ through the iterative processes

$$T_i^k \;:=\; T_{i+1}^{k-1} + \frac{T_{i+1}^{k-1} - T_i^{k-1}}{4^k-1}, \quad \begin{cases} k = 1,2,\ldots, \\ i = 0,1,\ldots, \end{cases} \tag{4.58}$$

and analogously

$$U_i^k \;:=\; U_{i+1}^{k-1} + \frac{U_{i+1}^{k-1} - U_i^{k-1}}{4^k-1}, \quad \begin{cases} k = 1,2,\ldots, \\ i = 0,1,\ldots, \end{cases} \tag{4.59}$$

then the error of $\{T_i^k\}$ as well as that of $\{U_i^k\}$ is of order $4^{-(k+1)i}$, and thus, for each k , the sequences of stage k converge faster than those of stage $(k-1)$.

This is Romberg's method for numerical quadrature: To compute the integral $\int_a^b f$ numerically, take the (rather crude) approximations T_i^0 and/or U_i^0 for $i = 0, 1, 2, \ldots$, and then apply the convergence accelerating processes (4.58) resp. (4.59).

Some authors denote only (4.58) as Romberg's method and call (4.59) (i.e. the use of the midpoint rule values) a *modified* Romberg method, since they might have overlooked that already Romberg himself introduced it.

Obviously, Processes (4.58) and (4.59) are nothing alse but the logarithmic extrapolation process (4.31) with $b = 2$ and $\rho_k = 2k$ for all k, and their application is justified by the existence of a logarithmic asymptotic expansion for the initial values of stage 0, as proved in Theorem 3.18 (cf. also Problem P4.9). Therefore we see that Romberg's method for numerical quadrature is a special case of the extrapolation process (4.31), and although it turned out later that the idea was not new, his paper was the one which initiated a tremendous lot of research and publications on the subject of extrapolation (see below and the Historical Notes), and most probably the book you hold in hands now never would have been written without Romberg's work (the decision if this would be a pity or not is up to you).

It is also pointed out already in W.ROMBERG [1] that the first extrapolations $\{T_n^1\}$ just reprocude Simpson's rule, while the second ones $\{T_n^2\}$ give the Newton-Cotes formula of order 6, denoted by Romberg as Gauß' rule, but nowadays better known as Milne's rule (cf. Problem P4.8). Moreover, Romberg pointed out that there is no Newton-Cotes formula which coincides with $\{T_n^k\}$ for $k \geq 3$.

The first rigorous analysis of Romberg's method and at the same time some generalizations of it were given in a series of papers by F.L.BAUER [1], F.L.BAUER, H.RUTISHAUSER & E.STIEFEL [1], H.RUTISHAUSER [1], H.RUTISHAUSER & E.STIEFEL [1], E.STIEFEL [1] and, independently, by R.BULIRSCH [1], R.BULIRSCH & J.STOER [1,3,4] in the early 60's. In particular, questions concerning error bounds and convergence proofs were considered in these papers. We shall investigate these topics in a more general context in Section 5.1, but provide here the following results in order to have together a rather complete description of Romberg's method in its classical form. However, the proofs we give in the following differ sometimes from the original ones considerably.

Let us first of all prove the convergence of all columns and diagonals in the Romberg tableau for, say, the $T-$ values. While it is clear by construction that that for each fixed $k(\leq r)$ we have

$$\lim_{i\to\infty} T_i^k = \int_a^b f(x)\,dx\;, \tag{4.60}$$

i.e. the columns in the Romberg tableau converge, the convergence of the diagonals is by no means trivial and must be proved. This is accomplished in the next theorem.

Theorem 4.11 (F.L.BAUER, H.RUTISHAUSER & E.STIEFEL [1], R.BULIRSCH [1]): *For each fixed* $i \in \mathbb{N}_0$ *we have*

$$\lim_{k\to\infty} T_i^k = \int_a^b f(x)\,dx\;. \tag{4.61}$$

Remark. Note that the statement of Theorem 4.11 describes the behaviour of the sequences $\{T_i^k\}$ as k tends to infinity, although we only had assumed $f \in C^{2r+1}[a,b]$, i.e. we can guarantee only the existence of a logarithmic asymptotic expansion of the finite order r. This means that the application of the extrapolation process (in this special case at least) makes also sense for $k > r$.

Proof of Theorem 4.11. First we note that each T_i^k is a linear combination of the initial values $T_i^0, \ldots, T_{i+k}^0$, i.e.

$$T_i^k = \sum_{\nu=i}^{i+k} L_{\nu i}^{(k)} \cdot T_\nu^0 \tag{4.62a}$$

with the Lagrange coefficients

$$L_{\nu i}^{(k)} = \prod_{\substack{\mu=i \\ \mu\neq\nu}}^{i+k} \frac{1}{1-4^{\mu-\nu}} , \tag{4.62b}$$

see (4.52). Since Relation (4.60) implies in particular that

$$\lim_{\nu\to\infty} T_\nu^0 = \int_a^b f(x)\,dx ,$$

the statement of Theorem 4.11 follows from Toeplitz' limit theorem (cf. e.g. K.Knopp [1]), if the following three conditions are satisfied:

$$(i) \qquad \sum_{\nu=i}^{i+k} L_{\nu i}^{(k)} = 1 \text{ for all } k ,$$

$$(ii) \qquad \lim_{k\to\infty} L_{\nu i}^{(k)} = 0 \text{ for each fixed } \nu ,$$

$$(iii) \qquad \sum_{\nu=i}^{i+k} |L_{\nu i}^{(k)}| \leq K \text{ for all } k ,$$

where K is some fixed constant.

But it is easily recognized that these conditions indeed hold, since (i) is a well-known property of the Lagrange numbers, while (ii) and (iii) can be read off from $(4.62b)$. □

With an a little bit more sophisticated proof it can be shown that the constant K in condition (iii) can be estimated by $K < 2$.

Moreover, generaliziations of Theorem 4.11 for other index sequences than Romberg's are to be found in R.Bulirsch & J.Stoer [1], R.Bulirsch [1] and P.-J.Laurent [1].

In our next theorem we will establish the exact error term of the extrapolates T_i^k ; this is not only an interesting result by itself, but it also will be a tool in order to derive an inclusion theorem and exit criteria for Romberg's method afterwards.

Theorem 4.12 (F.L.BAUER [1], R.BULIRSCH [1]): *For some $r \in I\!N_0$, consider a function $f \in C^{2r+2}[a,b]$, and the values T_i^k, computed according to Romberg's method (4.58). Then for each $k \leq r$, there exists a constant $R > 0$, independent of i and f, and a point $\xi \in (a,b)$, such that the relation*

$$T_i^k = \int_a^b f(x)\,dx + \frac{R}{4^{i(k+1)}} f^{(2k+2)}(\xi) \tag{4.63}$$

holds.

Proof. Let us analyze the error term in Euler's summation formula (Theorem 3.7), which is valid for each $k \leq r$. Using integration by parts and the properties of the Bernoulli polynomials and numbers as proved in Section 3.2, we find subsequently

$$\begin{aligned}
&\frac{1}{(2k+1)!}\int_0^n B_{2k+1}(t-[t])\cdot g^{(2k+1)}(t)\,dt \\
&\quad= \frac{1}{(2k+2)!}\Big\{\beta_{2k+2}\big(g^{(2k+1)}(n)-g^{(2k+1)}(0)\big) \\
&\qquad - \int_0^n B_{2k+2}(t-[t])\cdot g^{(2k+2)}(t)\,dt\Big\} \qquad (4.64)\\
&\quad= \frac{1}{(2k+2)!}\int_0^n \big(\beta_{2k+2}-B_{2k+2}(t-[t])\big)\cdot g^{(2k+2)}(t)dt.
\end{aligned}$$

Via Rolle's Theorem, we deduce from (3.18) and (3.23) that $(\beta_{2k+2} - B_{2k+2}(t-[t]))$ does not change its sign in $(0,n)$; hence there exists a point $\tilde{\xi} \in (0,n)$ such that the last line in (4.64) equals

$$\frac{1}{(2k+2)!} g^{(2k+2)}(\tilde{\xi}) \int_0^n \big(\beta_{2k+2}-B_{2k+2}(t-[t])\big)dt =$$

$$= \frac{n}{(2k+2)!} g^{(2k+2)}(\tilde{\xi})\,\beta_{2k+2}\,.$$

Applying now the transformation

$$g(t) \;=\; f(a + t(b-a)/n) \,,$$

we find that the error of the trapezoidal rule (4.55) for each $k \le r$ can be written in the form

$$T_{n_i}(f) \;=\; \int_a^b f(x)\,dx + \sum_{\mu=1}^{k} \frac{c_\mu}{4^{i\mu}} + \frac{C \cdot \beta_{2k+2}}{4^{i(k+1)}}\, f^{(2k+2)}(\xi) \qquad (4.65)$$

with

$$C \;:=\; \frac{(b-a)^{2k+3}}{(2k+2)!} \;>\; 0$$

and $\xi \in (a,b)$, cf. also the proof of Theorem 3.18.

After k extrapolation steps according to (4.58) we have eliminated the sum on the right hand side of (4.65) and obtain

$$T_i^k \;=\; \int_a^b f(x)\,dx \;+\; H_k \cdot \frac{C \cdot \beta_{2k+2}}{4^{i(k+1)}}\, f^{(2k+2)}(\xi) \,. \qquad (4.66)$$

with

$$H_k \;=\; \prod_{j=1}^{k} \frac{4^{j-k-1} - 1}{4^j - 1} \,. \qquad (4.67)$$

Since

$$\operatorname{sign}(H_k) = (-1)^k \quad \text{and} \quad \operatorname{sign}(\beta_{2k+2}) = (-1)^k \,, \qquad (4.68)$$

(cf. Problem P3.3), Theorem 4.12 is proved. □

As a direct consequence of Theorem 4.12 we obtain the following results on the monotonicity of the sequences $\{T_i^k\}$ and $\{U_i^k\}$, and on the bracketing of the true integral by these sequences.

Theorem 4.13 (J.ALBRECHT [1], T.STRÖM [2]): *Fix some natural number* $k \le r$ *and, in addition to the assumptions of Theorem 4.12, suppose that* $f^{(2k+2)}$ *does not change its sign in* (a,b), *say*

$$f^{(2k+2)}(\xi) \ \ge\ 0 \quad \text{for all} \quad \xi \in (a,b)\ . \tag{4.69}$$

Then the following inclusions hold for all $i \in \mathbb{N}_0$ *:*

$$U_i^k \ \le\ U_{i+1}^k \ \le\ \int_a^b f(x)\,dx \ \le\ T_{i+1}^k \ \le\ T_i^k\ . \tag{4.70}$$

Proof. The inequalities

$$\int_a^b f(x)\,dx \ \le\ T_{i+1}^k \ \le\ T_i^k$$

can be directly read off from Theorem 4.12.

To prove the relations concerning U_i^k, we first observe that an easy induction on k, starting with (4.57), yields the relation

$$T_{i+1}^k \ =\ \frac{T_i^k + U_i^k}{2}\ , \tag{4.71}$$

valid for all i. Thus, again by (4.63) we obtain

$$U_i^k \ =\ \int_a^b f(x)\,dx \ +\ \left(\frac{2}{4^{k+1}} - 1\right) \cdot \frac{R}{4^{i(k+1)}}\, f^{(2k+2)}(\xi)\,, \tag{4.72}$$

which already completes the proof of Theorem 4.13. □

T.STRÖM [1] had already proved in 1967 that from Condition (4.69) it follows

$$2\,T_{i+1}^k - T_i^k \ \le\ \int_a^b f(x)\,dx \ \le\ T_{i+1}^k\ , \tag{4.73}$$

which is, by (4.71), equivalent to

$$U_i^k \ \le\ \int_a^b f(x)\,dx \ \le\ T_{i+1}^k\ .$$

Moreover, in 1984 a generalization of Theorem 4.13 to the case of index sequences, which satisfy Condition (4.50), but are otherwise arbitrary, was found by K.JETTER [1].

The result of Theorem 4.13 provides at every step of Romberg's algorithms an inclusion of the true value and thus an exit criterion for these methods: If

$$|U_i^k - T_i^k| \; < \; \varepsilon$$

for some i, k and a prescribed accuracy ε, then each of these numbers can be taken as an approximation of the desired integral, and the corresponding error is guaranteed to be below ε. This can of course still be slightly improved by taking $\frac{1}{2}(U_i^k + T_i^k)$ as approximation, see T.HÅVIE [1] or T.STRÖM [1,2].

Another approach to the problem of obtaining exit criteria (or "stopping rules") is presented in the next theorem, which provides such criteria for the columns as well as the diagonals of the T-table; in particular, it should be emphasized that there is no need for calculating the values U_i^k.

Theorem 4.14 (K.-J.FÖRSTER [1,2]): *For all values of i and k, denote the errors appearing in the T-table by*

$$E_i^k \; := \; T_i^k \; - \int_a^b f(x)\,dx \; . \tag{4.74}$$

Then, in the the situation of Theorem 4.13, i.e. in particular if $f^{(2k+2)}(\xi) \geq 0$ for all $\xi \in (a,b)$, the following estimates are valid:

$$|E_{i+1}^k| \; \leq \; |T_{i+1}^k - T_i^k| \, , \qquad \textit{and} \tag{4.75}$$

$$|E_i^{k+1}| \; \leq \; \frac{4^{k+1} - 2}{4^{k+1}} \cdot |T_i^{k+1} - T_i^k| \, . \tag{4.76}$$

Proof. From (4.73) we obtain at once

$$2\,E_{i+1}^k - E_i^k \;\le\; 0 \;\le\; E_{i+1}^k\,, \tag{4.77}$$

hence

$$0 \;\le\; E_{i+1}^k \;\le\; E_i^k - E_{i+1}^k \;=\; T_i^k - T_{i+1}^k\,, \tag{4.78}$$

which proves (4.75) (cf. also H.BRASS [2], T.HÅVIE [3]).

To verify (4.76), we first observe that

$$\begin{aligned} E_{i+1}^k &= 4^{k+1} E_{i+1}^k - E_i^k - (4^{k+1}-2)E_i^{k+1} \\ &\le (4^{k+1}-2)(E_{i+1}^k - E_i^{k+1}) \\ &= (4^{k+1}-2)(T_{i+1}^k - T_i^{k+1})\,, \end{aligned} \tag{4.79}$$

where we have used (4.58) and (4.77).

Moreover, from (4.77) we obtain

$$E_i^{k+1} \;\ge\; E_i^{k+1} - E_{i+1}^k \;=\; T_i^{k+1} - T_{i+1}^k\,, \tag{4.80}$$

and from (4.58) again

$$4^{k+1}(T_i^{k+1} - T_{i+1}^k) \;=\; T_i^{k+1} - T_i^k\,. \tag{4.81}$$

Packing now together (4.79), (4.80) and (4.81), and keeping in mind that $E_i^{k+1} \ge 0$, we finally obtain the following chain of inequalities:

$$\begin{aligned} \frac{4^{k+1}-2}{4^{k+1}}\,(T_i^k - T_i^{k+1}) \;\ge\; E_i^{k+1} &\;\ge\; T_i^{k+1} - T_{i+1}^k \\ = \frac{1}{4^{k+1}}\,(T_i^{k+1} - T_i^k) &\;\ge\; \frac{4^{k+1}-2}{4^{k+1}}\,(T_i^{k+1} - T_i^k)\,, \end{aligned}$$

since $(T_i^{k+1} - T_{i+1}^k) \le 0$, due to (4.79). This completes the proof of Theorem 4.14. □

Generalizations of this result for extrapolation of arbitrary logarithmic asymptotic expansions, which satisfy a certain normality condition, were given in G.WALZ [4], see Section 5.2.

Much work on several generalizations of Romberg's original idea has been done in the meantime; in addition to the questions already mentioned above, people were interested mainly in problems concerning integrands with singularities (see Section 3.4), other index sequences than $\{n_i\} = \{2^i\}$ (R.BULIRSCH [1], R.BULIRSCH & J.STOER [1], A.MEIR & A.SHARMA [1]), multivariate integration (F.L.BAUER [1], A.SCHÖNHAGE [1], H.ANTES [1,2]), and other applications than quadrature (F.L. BAUER, H.RUTISHAUSER & E.STIEFEL [1], H.RUTISHAUSER [1]).

Much later, still another computational advantage of Romberg's method was found out, namely the fact that is is well-suited for the use on parallel computers, see C.CHIRUTA [1] or D.J.EVANS & G.M.MEGSON [1]).

4.3.2 Richardson Extrapolation and Gragg's Method. The numerical solution of differential equations has always been one of the main topics in numerical analysis and stimulated research in many branches of it; this is also true for the theory of asymptotic expansions and extrapolation methods.

In a certain sense, the story begins with the investigations of L. F. Richardson and his coworkers in the early years of our century: Using the results of some previous work, e.g. L.F.RICHARDSON [1,2], in 1927 the famous quasi-joint paper L.F.RICHARDSON [3] & J.A.GAUNT [1] on "the deferred approach to the limit" was published.

In the first part, Richardson investigates the h^2-extrapolation (nowadays called Richardson extrapolation) as a method of convergence acceleration for rather general discretization processes. To be more precise, suppose that we have a discretization method $T(h)$

with a certain "meshsize" $h > 0$, such that the convergence behaviour of this method can be described by

$$T(h) = c_0 + c_1 h^2 + O(h^4) \quad \text{for } h \to 0 . \tag{4.82}$$

Then we combine the approximations obtained with two different meshsizes h_0 and h_1 by setting

$$T(h_0, h_1) := \frac{h_1^2 T(h_0) - h_0^2 T(h_1)}{h_1^2 - h_0^2} , \tag{4.83}$$

and get a better approximation of c_0. This is the *method of h^2-extrapolation,* and of course it can be recovered as a special case of the extrapolation process (4.45) by choosing $k = 1$, $i = 0$, $\eta = 2$, and $n = 1/h$.

Richardson applied this method to a great variety of discretized problems, such as the solution of Laplace's equation in a square, computing Fourier coefficients, or solving a certain differential eigenvalue problem, cf. also D.C.JOYCE [1].

In J.A.GAUNT [1], the second part of this study, the author investigated in particular the numerical solution of an initial value problem for (a system of) first order ordinary differential equations by using the simplest linear multistep method of Nyström type, the midpoint rule (cf. (3.215)), denoted by himself as the "method of interpenetrating lattices".

Using the notation introduced in Section 3.6, Gaunt's approach and his main result can be stated as follows: To solve the initial value problem

$$y' = f(t, y) , \quad y(t_0) = y_0 \tag{4.84}$$

in a point $x > t_0$ numerically, choose a discretization parameter $n \in I\!N$, define the stepsize $h := (x - t_0)/n$, and compute

$$\left.\begin{aligned} y_1 &= \Phi(h) , \quad \text{and} \\ y_{\nu+2} &= y_\nu + 2hf(t_{\nu+1}, y_{\nu+1}) \quad \text{for } \nu = 0, \ldots, n-2 , \end{aligned}\right\} \tag{4.85}$$

where Φ is a suitable starting procedure. Then the asymptotic behaviour of the approximations $y_n := y(x,h)$ can be desribed by

$$y(x,h) \;=\; y(x) + c_1 h^2 + O(h^4) \quad \text{for } h \to 0\,, \tag{4.86}$$

i.e. a relation of the form (4.82) holds, and Richardson's h^2-extrapolation can be applied.

Interestingly, we find in J.A.GAUNT [1] the remark that "There would be no essential difficulty in extending the expansion [(4.86)] beyond the fourth power of h; but such a refinement would have little practical value." However, this might have been right in the year 1927, when all calculations had to be done by hand, but from a modern point of view one cannot agree with this opinion.

Gaunt still had some problems with finding a suitable starting procedure; he established certain formulae, in accordance with which the starting step had to be chosen. Thirty years later, R.DEVOGELAERE [1] re-proved and at the same time generalized Gaunt's result. In addition, he proposed to take a certain Runge-Kutta-type starting procedure, but also this was a little bit cumbersome and appears not to be appropriate with respect to the simplicity of the midpoint rule itself.

In view of all these difficulties, it was a big surprise and at the same time a great breakthrough, when W.B.GRAGG [1,2] was able to show that the simplest one-step method at all, namely Euler's method, can be taken as starting procedure Φ in (4.85). Since it was so important for the further development of extrapolation methods, let us state here this result as a theorem of itself (so-to-speak for historical reasons), although it can of course be regarded as a special case of our results in Section 3.6.

Theorem 4.15 (W.B.GRAGG [1,2]): *For the numerical solution of the initial value problem (4.84), apply the procedure*

$$\left.\begin{aligned} y_1 &= y_0 + h\,f(t_0, y_0)\,, \qquad \textit{and} \\ y_{\nu+2} &= y_\nu + 2hf(t_{\nu+1}, y_{\nu+1}) \quad \textit{for} \;\; \nu = 0, \ldots, n-2\,, \end{aligned}\right\} \tag{4.87}$$

i.e. a method of the type (4.85) with just one Euler step as starting procedure.

Then an asymptotic relation of the following type is valid for the sequence of final approximations $\{y_n\} = \{y(x,h)\}$ *:*

$$y(x,h) \;=\; y(x)+\sum_{\mu=1}^{N}\Big(d_{\mu 0}(x)+(-1)^n d_{\mu 1}(x)\Big)\,h^{2\mu}+o(h^{2N})\,. \quad (4.88)$$

The functions $d_{\mu 0}(x)$ *and* $d_{\mu 1}(x)$ *are independent of* h *, and the order* N *depends on the smoothness of* f *,*

Remark. Since the factor

$$(-1)^n \;=\; (-1)^{(x-t_0)/h} \quad (4.89)$$

depends on the stepsize h, Relation (4.88) does *not* establish a logarithmic asymptotic expansion in the sharp sense. However, if one restricts to subsequences of $\{y_n\}$ with even or odd indices only, then these subsequences indeed possess a logarithmic asymptotic expansion. Sometimes, this effect is expressed by saying that the method (4.87) produces two different asymptotic expansions, one for even and one for odd indices only.

Proof of Theorem 4.15. Since the method under consideration is of order 2 and satisfies the assumptions of Theorem 3.64, we readily obtain, after some obvious changes of notation, the representation

$$y(x,h) \;=\; y(x)+\sum_{\mu=2}^{m}\Big(e_{\mu 0}(x)+(-1)^n e_{\mu 1}(x)\Big)h^{\mu}+O(h^{m+1}) \quad (4.90)$$

with some integer m, depending on the smoothness of f.

Furthermore, since the method is symmetric, we may apply Theorem 3.66 and find that all terms on the right hand side of (4.90) with odd indices drop out, and finally setting $N := [m/2]$ and $d_{\mu j} = e_{2\mu,j}$ for $j = 0, 1$ completes the proof of Theorem 4.15. □

Based on this fundamental result, *Gragg's method* (W.B.GRAGG [1,2]) for the numerical solution of the initial value problem (4.84) can now be stated as follows:

Choose an increasing sequence of (preferably even) natural numbers $\{n_i\}$, and compute, using (4.87), the corresponding sequence of approximations $\{y_{n_i}\}$. Then apply a suitable extrapolation method in order to obtain faster converging sequences and thus good approximations of the true value $y(x)$.

In W.B.GRAGG [2] it is proposed to use the rational algorithm (4.49) from R.BULIRSCH & J.STOER [1] in order to manage the above-mentioned extrapolation. Therefore, Gragg's method is sometimes also denoted as *G(ragg)/B(ulirsch)/S(toer)-algorithm*, see also J.STOER [2].

Also, Gragg suggested to use local extrapolation instead of the global one; therefore, it is time to say a few words about the expressions "local extrapolation" and "global extrapolation ", which appear within the context of the numerical solution of ordinary differential equations.

By *global* extrapolation it is meant that one applies the discretization method under consideration – for several distinct values of h – always over the whole interval $[t_0, x]$, and then uses a suitable extrapolation method; therefore, everything we have mentioned so far falls into the class of global extrapolation methods.

In contrast to that, one can of course also carry out this process (i.e. applying the discretization method for several values of h and then extrapolating) in each subinterval $[t_\nu, t_{\nu+1}]$ separately; in other words, one goes from one point t_ν to the next one $t_{\nu+1}$ by subdividing this interval still more, and extrapolating the several resulting estimates for $y(t_{\nu+1})$ in order to get a quite good approximation $y_{\nu+1}$ there. This latter value is then taken as the starting point for the next step. This process is usually denoted as *local* extrapolation.

However, although these two approaches can lead in practice to quite different results, especially when dealing with stiff differential equations, the idea of extrapolation that stands behind them is the same in both cases, and therefore we do not want to make further remarks in this direction.

The importance and powerfulness of Gragg's method cannot be overestimated at all, since it combines in a wonderful manner the two fundamental principles of extrapolation, simplicity (in application) and efficiency: The initializing discretization process (4.87) is completely explicit and easy to implement, but nevertheless leads to an h^2-expansion and thus makes the use of the extrapolation process so efficient.

The oscillating property of the coefficient functions, induced by (4.89), might still cause a certain weak instability of the process (4.87). Therefore, Gragg suggested to complete this process by adding a certain final smoothing step, which removes the first weakly unstable component $((-1)^{(x-t_0)/h} d_{11})$ from the expansion (4.88). For completenes, we present this result in the following corollary, leaving some details of the proof to the reader.

Corollary 4.16 (W.B.GRAGG [2]): *Assume the situation of Theorem 4.15 to be given and extend the there-defined procedure (4.87) by adding a final smoothing step as follows:*

$$\left.\begin{aligned} y_1 &= y_0 + h\,f(t_0, y_0)\,, \\ y_{\nu+2} &= y_\nu + 2hf(t_{\nu+1}, y_{\nu+1}) \quad \text{for } \nu = 0, \dots, n-1\,, \quad \text{and} \\ s(x,h) &= (y_{n-1} + 2y_n + y_{n+1})/4\,. \end{aligned}\right\} \tag{4.91}$$

Then $s(x,h)$ satisfies the asymptotic relation

$$\left.\begin{aligned} s(x,h) &= y(x) + \big(d_{10}(x) + y''(x)/4\big)\, h^2 \\ &\quad + \sum_{\mu=2}^{N} \big(\tilde{d}_{\mu 0}(x) + (-1)^n \tilde{d}_{\mu 1}(x)\big)\, h^{2\mu} + O(h^{2N+2})\,. \end{aligned}\right\} \tag{4.92}$$

Sketch of proof. Expand the functions y_{n-1}, y_{n+1}, $d_{10}(t_{n-1})$, $d_{10}(t_{n+1})$, $d_{11}(t_{n-1})$, and $d_{11}(t_{n+1})$ in a Taylor series, and then use the definition of $s(x,h)$ in (4.91). □

However, it was pointed out in L.F.SHAMPINE & L.S.BACA [1] that in many cases the effect of this smoothing step might not be too big, since the extrapolation process annihilates the coefficients of the expansion anyway. Moreover, some global error estimates for extrapolation methods based on the procedure in (4.91) can be found in L.F.SHAMPINE & L.S.BACA [2].

4.3.3 Numerical Computation of the Matrix Exponential. The most important matrix function is the matrix exponential, well-defined (cf. Theorem 3.14) for each quadratic matrix A through the series expansion

$$\exp(A) \;:=\; \sum_{\nu=0}^{\infty} \frac{1}{\nu!}\, A^{\nu} \tag{4.93}$$

with $A^0 := I$. However, although this matrix function is so prominent, e.g. within the theory of ordinary differential systems, almost all existing procedures for their numerical computation work well only for a relatively small class of matrices A; therefore, it is no wonder that the study of C.MOLER & C.VANLOAN [1] bears the title "Nineteen *dubious* ways to compute the exponential of a matrix".

Later, in G.WALZ [1,2] another algorithm was proposed, which seems to be a little less dubious, in particular because there is no restriction at all concerning the input matrix A. This algorithm is nothing else but a special case of the extrapolation procedure (4.31), which can obviously also be applied to matrix sequences, since it is completely linear. The initial values of stage $k = 0$ are computed via the sequence of matrix functions

$$S_n(A) \;:=\; (I + \frac{1}{n}\, A)^n \,, \tag{4.94}$$

which possesses a logarithmic asymptotic expansion of the form

$$S_n(A) \;=\; \exp(A) + \sum_{\mu=1} \frac{1}{n^{\mu}} C_{\mu}(A) \tag{4.95}$$

with matrix functions C_μ, $\mu = 1, 2, \ldots$, see Subsections 3.3.2 and 3.6.1, in particular Lemma 3.45.

For completeness, we list here this algorithm for the numerical computation of $\exp(A)$ for an arbitrary quadratic matrix A:

1. With some user-chosen maximal index K, compute for $i = 0, \ldots, K$, the matrices

$$Y_i^0 := (I + 2^{-i} A)^{2^i} ;$$

2. For $k = 1, \ldots, K$ and $i = 0, \ldots, K - k$, compute

$$Y_i^k = Y_{i+1}^{k-1} + \frac{1}{2^k - 1} (Y_{i+1}^{k-1} - Y_i^{k-1}) ;$$

3. Use the matrix Y_0^K as approximation of $\exp(A)$.

Only as an illustration of this algorithm, let us approximate $\exp(A)$ for the matrix

$$A := \begin{pmatrix} 1 & 0 & 0 \\ 1 & 1 & 0 \\ 0 & 1 & 1 \end{pmatrix} .$$

With $K = 11$, our algorithm computes the approximation

$$\begin{pmatrix} 2.7182818284590452 & 0.0000000000000000 & 0.0000000000000000 \\ 2.7182818284590452 & 2.7182818284590452 & 0.0000000000000000 \\ 1.3591409142295226 & 2.7182818284590452 & 2.7182818284590452 \end{pmatrix},$$

which is correct up to 16 decimals.

For the results of some more numerical examples the reader is referred to G.WALZ [1] or G.F.THOMAS [1].

4.4 Problems

P 4.1. Consider the process of approximative geometric extrapolation (4.7) (in other words: the generalized Aitken process (4.8)).

Is the following statement true or false: Each of the sequences $\{\tilde{x}_n^{(k)}\}_{n\in\mathbb{N}}$ possesses a geometric asymptotic expansion of the form

$$\tilde{x}_n^{(k)} = c_0 + \sum_{\mu=k+1}^{m} \tilde{c}_\mu^{(k)} \lambda_\mu^n + o(\lambda_m^n) \quad \text{for } n \to \infty$$

with certain coefficients $\tilde{c}_\mu^{(k)}$ (cf. Equation (4.3)!)?

P 4.2. What do you think was the reason for assuming $N > m$ in Equation (4.10)?

P 4.3. Apply the exact extrapolation process for special geometric asymptotic expansions (4.9) with $\lambda = 1/2$ to the sequence $\{x_n\}$, defined by

$$x_0 = 2, \quad x_{n+1} = T(x_n) \qquad \text{for } n = 0, 1, \dots,$$

where T is the Newton iteration function from (4.14). Compare your results with those of the approximated extrapolation process, as shown in Table 4.4.

P 4.4. Give an explicit proof of Relation (4.71).

P 4.5. Consider a sequence of numbers $y_i^{(k)}$, defined according to the logarithmic extrapolation process (4.31), and the corresponding sequence of errors

$$e_i^{(k)} := y_i^{(k)} - c_0 .$$

Prove that for each fixed value of $k \le K$

$$\lim_{i\to\infty} \frac{e_i^{(k)}}{e_{i+1}^{(k)}} = b^{\rho_{k+1}} .$$

P 4.6. In addition to the index sequences introduced in Subsection 4.2.3, to be used in connection with the extrapolation process

(4.45), F.L.BAUER, H.RUTISHAUSER & E.STIEFEL [1] proposed to take

$$\{1, 2, 3, 6, 9, 18, 27, \ldots\}\ , \text{ i.e. } \quad n_i = \begin{cases} 3^{\frac{i}{2}}\,, & \text{if } i \text{ is even,} \\ 2 \cdot 3^{\frac{i-1}{2}}\,, & \text{if } i \text{ is odd.} \end{cases}$$

Write down explicitly the process (4.45) for this sequence.

P 4.7 (P.BJØRSTADT, G.DAHLQUIST & E.GROSSE [1], N.OSADA [1]). Consider a sequence $\{\sigma_n\}$ with the logarithmic asymptotic expansion

$$\sigma_n \;=\; c_0 \;+\; \sum_{\mu=1} \frac{c_\mu}{n^{\mu-1+\vartheta}}\,,$$

where ϑ is some positive real number.
Prove that for each fixed k the sequence $\{\sigma_n^{(k)}\}$, defined through the extrapolation process

$$\sigma_0^{(0)} := 0\,, \quad \sigma_n^{(0)} := \sigma_n \quad \text{for } n = 1, 2, \ldots,$$

$$\sigma_n^{(k+1)} \;:=\; \sigma_n^{(k)} - \frac{2k+1+\vartheta}{2k+\vartheta} \cdot \frac{(\sigma_{n+1}^{(k)} - \sigma_n^{(k)})(\sigma_n^{(k)} - \sigma_{n-1}^{(k)})}{\sigma_{n+1}^{(k)} - 2\sigma_n^{(k)} + \sigma_{n-1}^{(k)}}$$
$$\text{for } k = 0, 1, 2, \ldots \quad \text{and} \quad n = k+1, k+2, \ldots\,,$$

satisfies the asymptotic relation

$$\sigma_n^{(k)} \;=\; c_0 \;+\; O(n^{-(\vartheta+2k)}) \quad \text{for} \quad n \to \infty\,.$$

P 4.8. Prove that the sequences $\{T_i^1\}$ and $\{T_i^2\}$, defined through Romberg's process (4.58), just reproduce Simpson's rule respectively Milne's rule for numerical quadrature.

P 4.9. Using relation (4.57), give a one-line proof of the fact that the sequence of midpoint rule values $\{U_{n_i}(f)\}$ (cf. (4.56)) possesses a logarithmic asymptotic expansion.

P 4.10. Let p be an arbitrary even number. Combining some results from Subsections 3.6.2 and 4.3.2, prove that there exists an explicit Runge-Kutta method of order p.
Hint: Investigate the error term of the extrapolated midpoint rule.

5. Error Bounds, Stopping Rules and Monotonicity

So far, we have learned the fundamentals about extrapolation methods, have studied some of the standard examples for them, and, in particular, we have seen the extrapolation methods under consideration establish very easy convergence acceleration processes.

Now there arise questions concerning error bounds, exact ones as well as aymptotic ones, stopping rules for the extrapolation process, monotonicity preservation, error propagation and so on. These topics, some of which we have already scratched in connection with Romberg's method (Subsection 4.3.1), will be investigated in the present section; we will concentrate almost exclusively on the process of logarithmic extrapolation, which is surely the most widespread one, but remark here and at several places that many of the following results carry over to geometric extrapolation in an obvious way, see e.g. Problem P 5.1.

5.1 Asymptotic Error Bounds and Inclusions

Since extrapolation is a process which is based on the asymptotic behaviour of certain sequences, it is quite natural that the error bounds we will obtain for it are normally of asymptotic nature as well, i.e. typically the statements are of the type "for some k (or i) large enough, the following bound holds". However, in some cases it is also possible to derive, without imposing further conditions on the asymptotic expansions under consideration, exact (i.e. non-asymptotic) error bounds; we will begin this section with some remarks on such cases. Moreover, we will see in Section 5.2 how to impose reasonable normality conditions on the asymptotic expansions in order to obtain exact error bounds and also stopping rules for rather general classes of expansions.

5.1.1 Some Remarks on Exact Error Bounds. Probably the best understood and investigated extrapolation method is Romberg's one, which we studied in Subsection 4.3.1. There, we had seen that the error of the extrapolates T_i^k can be described exactly by the term

$$\frac{R}{4^{i(k+1)}}\, f^{(2k+2)}(\xi)\,,$$

see (4.63), and, moreover, that under the additional condition

$$f^{(2k+2)} \geq 0 \quad \text{in} \quad (a,b) \tag{5.1}$$

the sequences $\{T_i^k\}$ and $\{U_i^k\}$ *include* the true value of the integral (Theorem 4.13); also, stopping rules could be given (Theorem 4.14). Altogether we can say that for Romberg's method we are in the lucky situation to have all types of exact error bounds at hand – *but* of course most of them only under the condition that (5.1) holds, which may be hard to verify in practice!

T.Ström [3] and T.Ström & J.N.Lyness [1] derived results similar to the just mentioned ones for the method of numerical differentiation by extrapolation, i.e. for the application of the logarithmic extrapolation process (4.31) to the sequence of, say, central difference operators $\{\delta_n^c(g;m,x)\}$, which were proved in Theorem 3.35 to possess an logarithmic asymptotic expansion. But also in this case one has to assume that certain higher order derivatives of the function under consideration do not change sign.

Now, besides integration there is another field in numerical analysis where asymptotic expansions and extrapolation play an important role since long ago, namely the computation of the magic number π (see our Historical Notes), and we may hope to find at least here some results on exact error bounds.

Our candidate is Archimedes' method, which served as the very first introductory example for logarithmic asymptotic expansions in Subsection 3.1.1. Remember that Archimedes used the sequences

$$A_{2n} := n \cdot \sin\left(\frac{\pi}{n}\right) \quad \text{and} \quad U_n := n \cdot \tan\left(\frac{\pi}{n}\right), \tag{5.2}$$

which describe the area of a regular polygon, inscribed respectively circumscribed to the unit circle. It is clear from geometrical considerations that for each pair of natural numbers (n_1, n_2) the inclusion

$$A_{n_1} < \pi < U_{n_2} \tag{5.3}$$

is true. Moreover, $\{A_n\}$ as well as $\{U_n\}$ possesses a logarithmic asymptotic expansion with limit π and can be computed recursively, such that the process of logarithmic extrapolation can be applied, and the hope is that the inclusion (5.3) keeps valid for all extrapolated values; but unfortunately this is *false*, since already for the first extrapolates of U_n it turns out that

$$\frac{4\,U_{2n} - U_n}{3} < \pi . \tag{5.4}$$

This means that the extrapolations of the two Archimedean sequences $\{A_n\}$ and $\{U_n\}$ do *not* establish inclusions of the true value in each step. (To tell the whole truth: the entries of the U_i^k– table, i.e. the extrapolates of $\{U_n\}$, alternate in sign, as k increases, such that we indeed obtain an inclusion of π by these values alone; but to establish this, we need some results to be proved in Section 5.2; see there, in particular Subsection 5.2.3, also for some more details on the just-made remarks).

As a quintessence of these considerations we see that exact error bounds for extrapolation methods are something very, very rare, and therefore bounds of *asymptotic* type are the more appropriate ones here, in particular if we remember that extrapolation is a process which relies on the asymptotic behaviour of the initial sequences.

5.1.2 Asymptotic Error Bounds for Logarithmic Extrapolation. For convenience, let us first briefly re-state the extrapolation processes we want to talk about on the next pages: Starting with a sequence $\{\sigma_n\}$, which possesses a logarithmic asymptotic expansion of the usual type, say

$$\sigma_n \;=\; c_0 \;+\; \sum_{\mu=1}^{m} \frac{c_\mu}{n^{\rho_\mu}} + o(n^{-Re\rho_m}) \quad \text{for } n \to \infty \,,$$

we have essentially two choices for applying a suitable extrapolation process.

The first one is to restrict to the special index sequences of the form $n_i = n_0 \cdot b^i$, $n_0, b \in I\!N$, $b \geq 2$. Then there is no restriction at all concerning the exponents ρ_μ, and following Lemma 4.6 the extrapolation reads

$$y_i^{(k)} \;=\; \frac{b^{\rho_k} \cdot y_{i+1}^{(k-1)} - y_i^{(k-1)}}{b^{\rho_k} - 1} \tag{5.5}$$

with $y_i^{(0)} = \sigma_{n_i} = \sigma_{n_0 \cdot b^i}$; this will be adressed as *standard* case in the sequel.

Alternatively, we had seen in Subsection 4.2.3 that we can also admit arbitrary index sequences $\{n_i\}$, but then we have to impose the condition that all ρ_μ's are multiples of a fixed positive real number, say $\rho_\mu = \eta \cdot \mu$, i.e., the underlying logarithmic asymptotic expansion must be a special one. Then we had seen in Lemma 4.8 and Corollary 4.9 that the extrapolation process

$$Y_i^{(k)} \;=\; \frac{n_{i+k}^{\eta} Y_{i+1}^{(k-1)} - n_i^{\eta} Y_i^{(k-1)}}{n_{i+k}^{\eta} - n_i^{\eta}} \tag{5.6}$$

with $Y_i^{(0)} = \sigma_{n_i}$ is the right one; we will denote (5.6) as *modified (logarithmic) extrapolation* in this section.

In the following we present essentially two results on asymptotic error bounds and inclusions, each of which applies to both of the just mentioned cases.

In the first approach, due to R. Bulirsch & J. Stoer, one tries to recover the Romberg situation in the following sense: Starting with a given sequence of extrapolates $\{y_i^{(k)}\}$, one *defines* in a suitable way another sequence, say $\{u_i^{(k)}\}$, such that both sequences include the true value - at least asymptotically. The precise statement is as follows:

Theorem 5.1 (R. Bulirsch & J. Stoer [3,4]):
a) With fixed $k < m$, consider some sequence $\{y_i^{(k)}\}$, generated by the standard logarithmic extrapolation process (5.5), and define a corresponding new sequence $\{u_i^{(k)}\}$ by setting

$$u_i^{(k)} := (1+\alpha_k)\, y_{i+1}^{(k)} - \alpha_k\, y_i^{(k)} \tag{5.7}$$

for all i, where the numbers α_k themselves are defined through

$$\alpha_k := 1 + \frac{2}{b^{\rho_{k+1}} - 1}\,. \tag{5.8}$$

Then there exists a constant s_k, independent of n_i, such that for $n_i \to \infty$ the following asymptotic relations hold:

$$(i)\ y_i^{(k)} = c_0 + n_i^{-\rho_{k+1}}\left(s_k + O(n_i^{-(\rho_{k+2}-\rho_{k+1})})\right),\ \textit{and} \tag{5.9a}$$

$$(ii)\ u_i^{(k)} = c_0 + n_i^{-\rho_{k+1}}\left(-s_k + O(n_i^{-(\rho_{k+2}-\rho_{k+1})})\right). \tag{5.9b}$$

In particular, if $s_k \neq 0$, we have

$$\lim_{i\to\infty} \frac{y_i^{(k)} - c_0}{u_i^{(k)} - c_0} = -1\,, \tag{5.10}$$

i.e., for fixed k *and large enough values of* i *, the sequences* $\{y_i^{(k)}\}$ *and* $\{u_i^{(k)}\}$ *converge monotonically from opposite sides to the true value* c_0 .

b) In the case of modified logarithmic extrapolation, i.e. with $\{Y_i^{(k)}\}$ *defined through the process (5.6), we have to replace the definition of* α_k *in (5.8) by*

$$\alpha_k \;=\; \alpha_{k,i} \;=\; 1 + \frac{2}{\left(\frac{n_{i+k+1}}{n_i}\right)^{\eta} - 1} \,. \tag{5.11}$$

Then, again the new sequences $\{U_i^{(k)}\}$ *have to be defined according to the rule*

$$U_i^{(k)} \;:=\; (1+\alpha_{k,i})\, Y_{i+1}^{(k)} - \alpha_{k,i}\, Y_i^{(k)}\,, \tag{5.12}$$

and the result corresponding to (5.9) reads

$$(i)\ Y_i^{(k)} \;=\; c_0 + (n_i \cdots n_{i+k})^{-\eta}\big(\tilde{s}_k + O(n_i^{-\eta})\big), \tag{5.13a}$$

$$(ii)\ U_i^{(k)} \;=\; c_0 + (n_i \cdots n_{i+k})^{-\eta}\big(-\tilde{s}_k + O(n_i^{-\eta})\big) \tag{5.13b}$$

with some constant $\tilde{s}_k$.

In particular, Relation (5.10) and the resulting considerations remain true for the modified case.

Proof. Let us consider the standard case first. Here, we have the representation

$$y_i^{(k)} \;=\; c_0 \;+\; \sum_{\mu=k+1}^{m} \frac{c_\mu^{(k)}}{n_i^{\rho_\mu}} \;+\; o(n_i^{-Re\rho_m})$$

with

$$c_\mu^{(k)} \;=\; c_\mu \cdot \prod_{j=1}^{k} \frac{b^{\rho_j-\rho_\mu} - 1}{b^{\rho_j} - 1}\,,$$

(cf. Formula (4.32)), from which we can deduce the relation

$$y_i^{(k)} = c_0 + n_i^{-\rho_{k+1}} \cdot \prod_{j=1}^{k} \frac{b^{\rho_j - \rho_{k+1}} - 1}{b^{\rho_j} - 1} \cdot \cdot \left(c_{k+1} + O(n_i^{-(\rho_{k+2}-\rho_{k+1})})\right) \quad \text{for } n_i \to \infty\,, \tag{5.14}$$

see also R.Bulirsch & J.Stoer [1] or J.W.Schmidt [1].

Putting

$$s_k := c_{k+1} \cdot \prod_{j=1}^{k} \frac{b^{\rho_j - \rho_{k+1}} - 1}{b^{\rho_j} - 1}\,, \tag{5.15}$$

we recognize that (5.14) is already of the desired form; hence, Relation (5.9a) is proved.

To show (5.9b), we first note that from the definition of α_k in (5.8) we obtain, after some elementary calculations, the relation

$$(1 + \alpha_k)\, b^{-\rho_{k+1}} - \alpha_k = -1\,. \tag{5.16}$$

Then it follows

$$\begin{aligned} u_i^{(k)} &= (1+\alpha_k)\, y_{i+1}^{(k)} - \alpha_k\, y_i^{(k)} \\ &\overset{(5.9a)}{=} c_0 + (1+\alpha_k)\, n_{i+1}^{-\rho_{k+1}}\, s_k - \alpha_k\, n_i^{-\rho_{k+1}}\, s_k + O(n_i^{-\rho_{k+2}}) \\ &\overset{(5.16)}{=} c_0 - n_i^{-\rho_{k+1}}\, s_k + O(n_i^{-\rho_{k+2}}) \end{aligned} \tag{5.17}$$

i.e. Equation (5.9b).

Since the validity of (5.10) is obvious, the proof of the assertions concerning the standard case is complete.

The proof of the statements made under b) runs along the same lines as above, such that we restrict ourselves to some sketches:

First we have to prove (5.13a), i.e. the analogue of (5.9a). This can either be done directly, as in the proof of Theorem 4.10 (see R.BULIRSCH & J.STOER [1]), or, following an idea due to J.W.SCHMIDT [1], by first considering the "expansion"

$$\sigma_n := n^{-\eta\mu}, \qquad \mu \in \mathbb{N} \quad \text{fixed},$$

and then using the linearity of the extrapolation process; see Problem P 5.2 for more details.

To verify (5.13b), we first note that – by analogy with (5.16) – one obtains from (5.11) the relation

$$(1+\alpha_{k,i})\left(\frac{n_i}{n_{i+k+1}}\right)^{\eta} - \alpha_{k,i} = -1\,;$$

in combination with (5.13a), this implies in turn that

$$\begin{aligned} U_i^{(k)} &= c_0 + \left(n_i \cdots n_{i+k}\right)^{-\eta} \cdot \\ &\quad \cdot \left((1+\alpha_{k,i})\left(\frac{n_i}{n_{i+k+1}}\right)^{\eta} \tilde{s}_k - \alpha_{k,i}\,\tilde{s}_k + O(n_i^{-\eta})\right) \\ &= c_0 + \left(n_i \cdots n_{i+k}\right)^{-\eta}\left(-\tilde{s}_k + O(n_i^{-\eta})\right). \end{aligned}$$

This completes the proof of Theorem 5.1. □

Remark. If k is not too small, then the values of α_k resp. $\alpha_{k,i}$ will be very close to 1. This means that the easy-to-calculate sequences

$$\tilde{u}_i^{(k)} := 2\,y_{i+1}^{(k)} - y_i^{(k)} \qquad (5.18a)$$

resp.

$$\tilde{U}_i^{(k)} := 2\,Y_{i+1}^{(k)} - Y_i^{(k)} \qquad (5.18b)$$

may in practice serve as substitutes for the $u_i^{(k)}$'s resp. $U_i^{(k)}$'s themselves; cf. also Problem P 5.3 or R.BULIRSCH & J.STOER [3].

In order to illustrate the statements of Theorem 5.1, we apply the extrapolation process (5.5) with $n_0 = 1$ and $b = 2$ (which is the most popular choice) to the sequence $\sigma_n^e = (1 + 1/n)^n$, which we know (cf. Lemma 3.45) to possess a logarithmic asymptotic expansion of arbitrary order with $\rho_\mu = \mu$ for all μ and limit $c_0 = e = 2.718281828\ldots$.

Table 5.1 shows – in usual Romberg form – the resulting values of $y_i^{(k)}$ (top) and $u_i^{(k)}$ (bottom) for $k = 0, \ldots, 4$ and $i = 0, \ldots, 6-k$; it is obvious that *each* pair $(y_i^{(k)}, u_i^{(k)})$ includes the true value e.

2.0000000 3.0000000				
	2.5000000 2.8541667			
2.2500000 3.0156250		2.6770833 2.7506746		
	2.6328125 2.7857466		2.7138790 2.7221807	
2.4414063 2.9389193		2.7092795 2.7262613		2.7180298 2.7185189
	2.6901628 2.7432553		2.7177704 2.7187630	
2.5657845 2.8543604		2.7167090 2.7197623		2.7182667 2.7182965
	2.7100725 2.7260172		2.7182357 2.7183266	
2.6379285 2.7941750		2.7180448 2.7185117		2.7182811 2.7182825
	2.7160518 2.7204465		2.7182783 2.7182853	
2.6769901 2.7584094		2.7182491 2.7183141		
	2.7176998 2.7188553			
2.6973449 2.7389212				

Table 5.1. Illustration of Theorem 5.1: Values $y_i^{(k)}$ (top) and $u_i^{(k)}$ (bottom)

With almost the same assumptions and case distinctions as in Theorem 5.1, one can also prove still another type of asymptotic inclusion theorem; in contrast to the $u_i^{(k)}$'s from above, one now defines in a *non-linear* manner new sequences $v_i^{(k)}$ in order to obtain an inclusion for the desired value c_0.

Theorem 5.2 (J.W.SCHMIDT [1]): *Consider some sequence* $\{y_i^{(k)}\}$, *generated either by the standard logarithmic extrapolation process (5.5), or by the modified extrapolation process (5.6), and define in each case a new sequence* $\{v_i^{(k)}\}$ *by setting*

$$v_i^{(k)} := y_{i+1}^{(k-1)} + \frac{y_i^{(k)} - y_{i+1}^{(k-1)}}{1 - \dfrac{y_i^{(k)} - y_{i-1}^{(k)}}{y_{i+1}^{(k-1)} - y_i^{(k-1)}}} \tag{5.19}$$

for all $i \geq 1$. *Then*

$$\lim_{i\to\infty} v_i^{(k)} = c_0 .$$

Moreover, if the coefficients in the underlying asymptotic expansion satisfy $c_k\, c_{k+1} \neq 0$, *for all* i *large enough the relation*

$$\left(y_i^{(k)} - c_0\right) \cdot \left(v_i^{(k)} - c_0\right) < 0 , \tag{5.20}$$

holds, i.e., the numbers $y_i^{(k)}$ *and* $v_i^{(k)}$ *include the true value* c_0.

Proof. Again we work out the proof for the case of standard extrapolation and refer the reader for the (analogous) proof of the modified case to the original paper J.W.SCHMIDT [1] or to his/her own considerations.

Let us first re-write $v_i^{(k)}$ in the form

$$v_i^{(k)} = \frac{y_{i+1}^{(k-1)} y_{i-1}^{(k)} - y_i^{(k)} y_i^{(k-1)}}{\left(y_{i+1}^{(k-1)} - y_i^{(k-1)}\right) - \left(y_i^{(k)} - y_{i-1}^{(k)}\right)} . \tag{5.21}$$

Next we use Relation (5.14), which we state here in the form

$$y_i^{(k)} = c_0 + n_i^{-\rho_{k+1}} \cdot \prod_{j=1}^{k} p_{k,j} \cdot \left(c_{k+1} + o(1)\right) \tag{5.22a}$$

for $n_i \to \infty$, where

$$p_{k,j} := \frac{b^{\rho_j - \rho_{k+1}} - 1}{b^{\rho_j} - 1}. \tag{5.22b}$$

Now the reader is encouraged to take some time and several sheets of blank paper: We insert (5.22) into the representation (5.21) and analyze one after another the different terms arizing there. Thus we find the numerator to be of the form

$$\begin{aligned} y_{i+1}^{(k-1)} y_{i-1}^{(k)} - y_i^{(k)} y_i^{(k-1)} &= \\ = c_0 \Big(n_{i+1}^{-\rho_k} \prod_{j=1}^{k-1} p_{k-1,j}(c_k + o(1)) &+ n_{i-1}^{-\rho_{k+1}} \prod_{j=1}^{k} p_{k,j}(c_{k+1} + o(1)) \\ - n_i^{-\rho_{k+1}} \prod_{j=1}^{k} p_{k,j}(c_{k+1} + o(1)) &- n_i^{-\rho_k} \prod_{j=1}^{k-1} p_{k-1,j}(c_k + o(1)) \Big) \\ + n_{i+1}^{-\rho_k} n_{i-1}^{-\rho_{k+1}} \prod_{j=1}^{k-1} p_{k-1,j} \prod_{j=1}^{k} p_{k,j} &\cdot (c_k + o(1))(c_{k+1} + o(1)) \\ - n_i^{-\rho_{k+1}} n_i^{-\rho_k} \prod_{j=1}^{k} p_{k,j} \prod_{j=1}^{k-1} p_{k-1,j} &\cdot (c_k + o(1))(c_{k+1} + o(1)). \end{aligned} \tag{5.23}$$

Analogously, the denominator of the quotient in (5.21) can be written as

$$\begin{aligned} (y_{i+1}^{(k-1)} - y_i^{(k-1)}) - (y_i^{(k)} - y_{i-1}^{(k)}) &= \\ = \left(n_{i+1}^{-\rho_k} - n_i^{-\rho_k}\right) \prod_{j=1}^{k-1} p_{k-1,j} &\cdot (c_k + o(1)) \\ - \left(n_i^{-\rho_{k+1}} - n_{i-1}^{-\rho_{k+1}}\right) \prod_{j=1}^{k} p_{k,j} &\cdot (c_{k+1} + o(1)). \end{aligned} \tag{5.24}$$

One recognizes that the coefficient of c_0 in (5.23) cancels out, such that we are left to analyze the quotient of the remaining term in (5.23) and the denominator (5.24).

Again we consider first the numerator term, i.e. the last two lines of (5.23), and find that it can be brought into the form

$$(b^{\rho_{k+1}-\rho_k}-1)\cdot n_i^{-(\rho_{k+1}+\rho_k)}\prod_{j=1}^{k}(p_{k-1,j}p_{k,j})\cdot\big(c_k c_{k+1}+o(1)\big)\,, \tag{5.25}$$

while the denominator term can be simplified once more to

$$\begin{aligned} &(b^{-\rho_k}-1)\cdot n_i^{-\rho_k}\prod_{j=1}^{k-1}p_{k-1,j}\cdot(c_k+o(1)) \\ &-n_i^{\rho_k-\rho_{k+1}}\cdot(1-b^{\rho_{k+1}})\cdot n_i^{-\rho_k}\prod_{j=1}^{k}p_{k,j}\cdot(c_{k+1}+o(1)) \\ &=(b^{-\rho_k}-1)\cdot n_i^{-\rho_k}\prod_{j=1}^{k}p_{k-1,j}\cdot(c_k+o(1))\,. \end{aligned} \tag{5.26}$$

Inserting now Relations (5.25) and (5.26) into (5.21), we finally obtain

$$v_i^{(k)}=c_0+\frac{b^{\rho_{k+1}-\rho_k}-1}{b^{-\rho_k}-1}\cdot n_i^{-\rho_{k+1}}\cdot\prod_{j=1}^{k}p_{k,j}\cdot\big(c_{k+1}+o(1)\big)\,. \tag{5.27}$$

Now, if i is large enough such that the $o(1)-$ terms have no influence on the signs of the expressions in (5.22) and (5.27), a comparison of these two relations proves the statement of Theorem 5.2, since

$$\frac{b^{\rho_{k+1}-\rho_k}-1}{b^{-\rho_k}-1}<0 \tag{5.28}$$

for all k. □

We want to illustrate also the statement of Theorem 5.2 by a numerical example, in fact by the same one as after Theorem 5.1 (see there for details). The entries of the corresponding Table 5.2 are therefore the values of $y_i^{(k)}$ (top) and $v_i^{(k)}$ (bottom). Once again, inclusion of the true value $e = 2.718281828\ldots$ is given for each pair $(y_i^{(k)}, v_i^{(k)})$.

2.0000000				
	2.5000000			
2.2500000		2.6770833		
	2.6328125 3.0666667		2.7138790	
2.4414063		2.7092795 2.7337482		2.7180298
	2.6901628 2.7965829		2.7177704 2.7189378	
2.5657845		2.7167090 2.7206598		2.7182667 2.7182989
	2.7100725 2.7375710		2.7182357 2.7183377	
2.6379285		2.7180448 2.7186182		2.7182811 2.7182826
	2.7160518 2.7231117		2.7182783 2.7182860	
2.6769901		2.7182491 2.7183268		
	2.7176998 2.7194930			
2.6973449				

Table 5.2. Illustration of Theorem 5.2: Values $y_i^{(k)}$ (top) and $v_i^{(k)}$ (bottom)

Remark. (i) The result of Theorem 5.2 remains true if one replaces $v_i^{(k)}$ from (5.19) by

$$\tilde{v}_i^{(k)} := y_{i+1}^{(k-1)} + \frac{y_i^{(k)} - y_{i+1}^{(k-1)}}{1 - \dfrac{y_{i+1}^{(k)} - y_i^{(k)}}{y_{i+2}^{(k-1)} - y_{i+1}^{(k-1)}}}\,; \tag{5.29}$$

this can be proved with exactly the same methods as above, see the original paper.
(ii) It is also worked out in J.W.SCHMIDT [1] that the $v_i^{(k)}$'s resp. the $\tilde{v}_i^{(k)}$'s may serve as asymptotic inclusions for the result of Aitken's Δ^2- process, which is nothing else but the first step of the approximative geometric extrapolation process (4.8), and also of the $\epsilon-$ algorithm, cf. Subsection 4.1.1. Moreover, results similar to those of Theorems 5.1 and 5.2 for the case of logarithmic asymptotic expansions with exponents of the form

$$\rho_\mu = \rho + (\mu - 1)\eta,$$

($\rho \neq \eta$ in general) are derived in J.W. SCHMIDT [2].

5.2 Error Bounds and Stopping Rules for Normal Expansions

If we want to have error bounds or stopping rules, which are exact in the sense that there is no condition like "for i large enough" in the statements, we must look for some more information on the structure of the underlying asymptotic expansions.

One possibility is to ask for the *sign behaviour* of the coefficients c_μ in the asymptotic expansion

$$\sigma_n = c_0 + \sum_{\mu=1}^{m} \frac{c_\mu}{n^{\rho_\mu}} + o(n^{-Re\rho_m}). \tag{5.30}$$

The following normality condition on this behaviour has turned out to be weak enough on one hand, such that it is satisfied still by a great class of asymptotic expansions, and strong enough on the other hand, such that we can prove some results on error bounds for such asymptotic expansions. As before, we will concentrate on logarithmic extrapolation and, moreover, we restrict ourselves now on the standard case and choose the index sequences $n_i = b^i$, but impose no further restrictions on the exponents ρ_μ .

We briefly recall from Lemma 4.6 that in this case each of the sequences $\{y_i^{(k)}\}_{i\in\mathbb{N}}$ possesses a logarithmic asymptotic expansion of the form

$$y_i^{(k)} = c_0 + \sum_{\mu=k+1}^{m} \frac{c_\mu^{(k)}}{n_i^{\rho_\mu}} + o(n_i^{-Re\rho_m}) \quad \text{for } n_i \to \infty \qquad (5.31a)$$

with the coefficients

$$c_\mu^{(k)} = c_\mu \cdot \prod_{j=1}^{k} \frac{b^{\rho_j-\rho_\mu}-1}{b^{\rho_j}-1} := c_\mu \cdot h_\mu^{(k)} . \qquad (5.31b)$$

With these notations and facts recalled, we can now state the definition of normality.

Definition 5.3: With some integer $k_0 \geq 0$ let there be given an asymptotic expansion of type (5.30) such that $\rho_{k_0+1} \geq 1$. This expansion is called k_0- *normal* (or just *normal*, if there is no doubt about k_0), if for all $m \geq \mu > k_0$ the coefficients c_μ are of either of the following forms:

They are all of the same sign; without any loss of generality we then assume

$$c_\mu > 0 \quad \text{for all } \mu > k_0 , \qquad (5.32)$$

and call the asymptotic expansion to be normal of *Type A*.

They alternate in sign, say

$$(-1)^{\mu+1} c_\mu > 0 \quad \text{for all } \mu > k_0 , \qquad (5.33)$$

and, in addition,

$$|c_\mu^{(k)}| > |c_{\mu+1}^{(k)}| \qquad (5.34)$$

holds for all $\mu > k \geq k_0$; then the asymptotic expansion is normal of *Type B*.

Without any doubt, normality of Type A is easier to handle than that of Type B (and has also other not so obvious advantages, as it will turn out later), since in particular condition (5.34) may be sometimes hard to verify in practice. However, examples at the end of this subsection will illustrate that there exists indeed a lot of non-trivial asymptotic expansions, which can easily be verified to be normal of Type B. Moreover, the results and in particular the proofs of these on the following pages will clearly indicate some possibilities for modifying condition (5.34) in particular cases.

Also, it should be remarked that these normality conditions are not as restrictive as they might seem at first sight; for example, most of the asymptotic expansions which occur in connection with the approximation of functions (cf. Section 3.3) and with numerical differentiation (cf. Section 3.5) are indeed normal, and even more, they are normal with $k_0 = 0$.

Similarly, Euler's summation formula provides some information on the normality of the asymptotic expansions appearing in connection with numerical quadrature, depending on the sign behaviour of the function's derivatives, and also in the context of computing π it will turn out that the resulting asymptotic expansions share the property of being normal, the type of normality still depending on the approach to be taken; we will elaborate this in Subsection 5.2.3.

Let us begin with an easy lemma on the behaviour of the extrapolated coefficients $c_\mu^{(k)}$ in the case of normal expansions:

Lemma 5.4: *Let $\{\sigma_n\}$ possess a normal asymptotic expansion with coefficients $c_\mu^{(k)}$. Then for $k_0 \leq k \leq m$ and $k+1 < \mu \leq m$ we have*

$$|c_\mu^{(k+1)}| < |c_\mu^{(k)}| . \tag{5.35}$$

This means that each step of the extrapolation procedure not only eliminates one coefficient of the expansion, but also diminishes all others.

To *prove* this lemma we just apply (5.31b) and use the fact that for normal expansions we have $b^{\rho_{k+1}} - 1 \geq 1$ for $k_0 \leq k$. □

On the next pages we will establish error bounds and stopping rules for both types of normal asymptotic expansions. Most of these results were originally worked out in G. WALZ [4], for some special cases, in particular Romberg's method, one may also have a look into the papers by K.-J. FÖRSTER [1] or H. NEUMANN [1].

5.2.1 A-Priori Error Bounds. Let us again denote by

$$e_i^{(k)} := y_i^{(k)} - c_0 \tag{5.36}$$

the *approximation error* of $y_i^{(k)}$ for all i and k. In this subsection we are interested in a–priori bounds for this error; such bounds are used in practice to choose the maximal number K of extrapolation stages in a suitable way.

In the case of normal expansions, we will derive such bounds from the following Theorem 5.5, which itself is of some importance:

Theorem 5.5 (G. WALZ [4]): *Let the sequence $\{\sigma_n\}$ possess a normal asymptotic expansion. Then for each k with $k_0 \leq k \leq m$ and $i \in I\!N$, the following statements on the sign of the approximation errors $e_i^{(k)}$ hold:*

$$(i)\ \ \operatorname{sign}(e_i^{(k)}) = (-1)^k \cdot \operatorname{sign}(c_{k+1}), \quad \textit{and} \tag{5.37}$$

$$(ii)\ \ \operatorname{sign}(e_i^{(k)}) = \begin{cases} (-1)^k, & \textit{if } \sigma_n \textit{ is of Type A,} \\ 1, & \textit{if } \sigma_n \textit{ is of Type B,} \end{cases} \tag{5.38}$$

Proof. From relations (5.31) we obtain at once the following explicit form of $e_i^{(k)}$:

$$\begin{aligned} e_i^{(k)} &= \frac{c_{k+1}}{b^{\rho_{k+1}\, i}}\, h_{k+1}^{(k)} + \frac{c_{k+2}}{b^{\rho_{k+2}\, i}}\, h_{k+2}^{(k)} + \frac{c_{k+3}}{b^{\rho_{k+3}\, i}}\, h_{k+3}^{(k)} + \cdots \\ &= \frac{c_{k+1}^{(k)}}{b^{\rho_{k+1}\, i}} + \frac{c_{k+2}^{(k)}}{b^{\rho_{k+2}\, i}} + \frac{c_{k+3}^{(k)}}{b^{\rho_{k+3}\, i}} + \cdots . \end{aligned} \tag{5.39}$$

Moreover, it is easily recognized that for all $\mu \geq k+1$ the $h_\mu^{(k)}$'s in (5.31b) satisfy

$$\operatorname{sign}(h_\mu^{(k)}) = (-1)^k , \tag{5.40}$$

i.e., for fixed k , all $h_\mu^{(k)}$ are of the same sign, while for fixed $\mu \geq k+2$, $h_\mu^{(k)}$ and $h_\mu^{(k+1)}$ are of opposite sign.

Recalling now our definition of normality and using the sign rules for alternating series (see e.g. K.KNOPP [1]), both statements of Theorem 5.5 now follow easily. □

Theorem 5.5 may be interpreted as follows: If the errors $e_i^{(k)}$ are displayed according to the Romberg tableau (Table 4.1), then, if $k \geq k_0$, the sign of $e_i^{(k)}$ does not change if we go through a fixed column of the tableau. If we go through the scheme in horizontal direction, then the sign of $e_i^{(k)}$ is either *constant* (normality of Type B), or *alternating* (Type A).

The proof of Theorem 5.5 also indicates how to modify condition (5.34) suitably in specific cases; for example, we could demand that

$$h_\mu^{(k)} \cdot b^{-\rho_\mu} \tag{5.41}$$

decreases fast enough, as μ increases.

Moreover, from Equation (5.39) it is obvious that if (5.33) holds, but (5.34) is not satisfied, then nevertheless the corresponding asymptotic expansion "becomes" normal, if i is large enough; but this of course leads again to a statement of asymptotic nature.

Now, a first and very general approach to the desired a-priori bounds could be the application of the triangle inequality to the expansion (5.39) of $e_i^{(k)}$, which in some cases leads to surprisingly good results, see e.g. G.MEINARDUS & G.MERZ [2].

But in the presently considered case of *normal* expansions, we can by straightforward computations derive the following bounds, which are in most cases better than those mentioned above and also easier to obtain:

Adopt the notation from above; if the asymptotic expansion is of *Type B,* we have for all i and $k_0 \le k \le m$:

$$\frac{c_{k+1}}{b^{\rho_{k+1} i}}\, h_{k+1}^{(k)} + \frac{c_{k+2}}{b^{\rho_{k+2} i}}\, h_{k+2}^{(k)} \le e_i^{(k)} \le \frac{c_{k+1}}{b^{\rho_{k+1} i}}\, h_{k+1}^{(k)}, \tag{5.42}$$

and, if it is of *Type A,* we have

$$\frac{c_{k+1}}{b^{\rho_{k+1} i}} |h_{k+1}^{(k)}| \le |e_i^{(k)}| < \frac{1}{\prod_{j=1}^{k}(b^{\rho_j} - 1)} \cdot \sum_{\mu=k+1}^{\infty} \frac{c_\mu}{b^{\rho_\mu i}}, \tag{5.43}$$

provided that the order m is arbitrary.

Moreover, if the exponents ρ_μ are, for example, of the form $\rho_\mu = \eta \cdot \mu$ (i.e., if the asymptotic expansion is a special one in the sense of Definition 3.40), and if the coefficients c_μ can be estimated from above by a constant $C \ge c_\mu$, then for $i > 0$ the right hand side of (5.43) simplifies to

$$|e_i^{(k)}| < \frac{C}{\prod_{j=1}^{k}(b^{\eta j} - 1)} \cdot \frac{b^{-i\eta(k+1)}}{1 - b^{-i\eta}}.$$

5.2.2 Stopping Rules. The a-priori error bounds given in the previous subsection can be used to estimate the number K of extrapolation stages that have to be examined in the worst case. This value of K can now be stored in the computer, and so the desired accuracy of the approximation can be guaranteed a priori.

Now it is of course not always necessary to compute all the $y_i^{(k)}$'s up to $k = K$, because in many cases the desired precision will already be reached by some $y_i^{(k)}$ with k significantly less than K.

So the question is: Do there exist methods, which check *during the extrapolation process,* if the desired accuracy is reached? For normal expansions this is in fact true, and in the following we will give for each of the both types of normal expansions such a stopping rule. The criterion to be used is the difference of the two elements $y_i^{(k)}$ and $y_j^{(k+1)}$, or, in other words, the improvement of the approximation one obtains by stepping from one column of the Romberg tableau to the next one.

For Type A expansions, a stopping rule is quite easy to obtain:

Theorem 5.6 (G.WALZ [4]): *Let the sequence $\{\sigma_n\}$ possess a normal asymptotic expansion of Type A and adopt the notations from above. Then*

$$\left|e_i^{(k+1)}\right| \le \left|y_i^{(k+1)} - y_j^{(k)}\right| \tag{5.44}$$

for $k_0 \le k \le m$ and all $i, j \in I\!N$.

Proof. First assume that k is even. Then, due to Theorem 5.5, we have $e_j^{(k)} \ge 0$ and $e_i^{(k+1)} \le 0$ for all $i, j \in I\!N$, hence

$$y_i^{(k+1)} \le c_0 \le y_j^{(k)} .$$

From this we immediately obtain

$$0 \le c_0 - y_i^{(k+1)} \le y_j^{(k)} - y_i^{(k+1)} ,$$

and because $c_0 - y_i^{(k+1)} = -e_i^{(k+1)} \ge 0$, the proof is complete.

For odd values of k the assertion can be proved in an analogous way. □

If $\{\sigma_n\}$ is of Type B, the case is a bit more complicated; here we have to make the additional assumption that for all i and $k \geq k_0$ there exists a real number $\alpha_i^{(k)} > 1$ such that

$$\alpha_i^{(k)} \cdot e_{i+1}^{(k)} \leq e_i^{(k)} . \tag{5.45}$$

We will show later how to choose $\alpha_i^{(k)}$ in a satisfactory manner; first we prove the following stopping criterion for Type B expansions.

Theorem 5.7 (G.Walz [4]): *Let the sequence $\{\sigma_n\}$ possess a normal asymptotic expansion of Type B and adopt the notations from above. Furthermore, let $\alpha_i^{(k)}$ be a real number greater than 1 and satisfying (5.45). Then*

$$e_i^{(k+1)} \leq \frac{b^{\rho_{k+1}} - \alpha_i^{(k)}}{(\alpha_i^{(k)} - 1) \cdot b^{\rho_{k+1}}} \cdot (y_i^{(k)} - y_i^{(k+1)}) \tag{5.46}$$

for $k_0 \leq k \leq m$ and $i \in I\!N$.

Proof. Due to the linearity of the extrapolation process we have

$$e_i^{(k+1)} = \frac{b^{\rho_{k+1}} e_{i+1}^{(k)} - e_i^{(k)}}{b^{\rho_{k+1}} - 1} , \tag{5.47}$$

hence

$$0 = b^{\rho_{k+1}} \cdot (e_{i+1}^{(k)} - e_i^{(k+1)}) - e_i^{(k)} + e_i^{(k+1)} .$$

Using Condition (5.45), this implies in turn that

$$\begin{aligned}
(\alpha_i^{(k)} - 1) \cdot e_i^{(k+1)} &= \\
&= b^{\rho_{k+1}} \cdot (e_{i+1}^{(k)} - e_i^{(k+1)}) - e_i^{(k)} + \alpha_i^{(k)} e_i^{(k+1)} \\
&\leq b^{\rho_{k+1}} \cdot (e_{i+1}^{(k)} - e_i^{(k+1)}) + \alpha_i^{(k)} (e_i^{(k+1)} - e_{i+1}^{(k)}) .
\end{aligned}$$

Dividing by $(\alpha_i^{(k)} - 1)$, we obtain

$$\begin{aligned} e_i^{(k+1)} &\leq \frac{b^{\rho_{k+1}} - \alpha_i^{(k)}}{\alpha_i^{(k)} - 1} \cdot (e_{i+1}^{(k)} - e_i^{(k+1)}) \\ &= \frac{b^{\rho_{k+1}} - \alpha_i^{(k)}}{\alpha_i^{(k)} - 1} \cdot (y_{i+1}^{(k)} - y_i^{(k+1)}) , \end{aligned} \tag{5.48}$$

and since

$$y_{i+1}^{(k)} - y_i^{(k+1)} = \frac{y_i^{(k)} - y_i^{(k+1)}}{b^{\rho_{k+1}}} ,$$

inequality (5.48) proves the theorem. □

The determination of suitable numbers $\alpha_i^{(k)}$ is a problem of itself. In the (very) special case of integrating a function f with $f^{(2k+2)} \geq 0$ by Romberg's method, the results in Subsection 4.3.1 imply that $\alpha_i^{(k)}$ can be chosen equal to 2 for all i and k. Unfortunately, this does not carry over to arbitrary asymptotic expansions; in the sequel we therefore suggest two other possibilities for a suitable choice of $\alpha_i^{(k)}$.

The first one is based on the asymptotic structure of the extrapolation method; we note that for fixed k

$$\lim_{i \to \infty} \frac{e_i^{(k)}}{e_{i+1}^{(k)}} = b^{\rho_{k+1}} \tag{5.49}$$

(see Problem P 4.5), a limit which is usually almost reached after a few steps, and so a reasonable choice would be

$$\alpha_i^{(k)} := b^{\rho_{k+1}/2} . \tag{5.50}$$

If one doesn't like bounds which are only asymptotically correct, we can also define $\alpha_i^{(k)}$ in a way that (5.45) holds for all values of

i and k; to this end we note that, due to (5.42),

$$\frac{e_i^{(k)}}{e_{i+1}^{(k)}} \geq \left(\frac{c_{k+1}}{b^{\rho_{k+1} i}} \cdot h_{k+1}^{(k)} + \frac{c_{k+2}}{b^{\rho_{k+2} i}} \cdot h_{k+2}^{(k)}\right) / \left(\frac{c_{k+1}}{b^{\rho_{k+1}(i+1)}} \cdot h_{k+1}^{(k)}\right)$$

$$= b^{\rho_{k+1}} \cdot \left(1 + \frac{c_{k+2} h_{k+2}^{(k)}}{c_{k+1} h_{k+1}^{(k)}} \cdot b^{i(\rho_{k+1} - \rho_{k+2})}\right) .$$

In view of this, we may choose

$$\alpha_i^{(k)} := b^{\rho_{k+1}} \cdot \left(1 + \frac{c_{k+2} h_{k+2}^{(k)}}{c_{k+1} h_{k+1}^{(k)}} \cdot b^{i(\rho_{k+1} - \rho_{k+2})}\right) . \tag{5.51}$$

The additional amount of work caused by the use of (5.51) during the extrapolation process is not great, because one has to compute only once the numbers $(c_{k+2} h_{k+2}^{(k)})/(c_{k+1} h_{k+1}^{(k)})$, $b^{\rho_{k+1}}$ and $b^{\rho_{k+2}}$ for all k under consideration; the evaluation of $\alpha_i^{(k)}$ then needs only a few operations.

Nevertheles, the additional computational cost induced by the application of our stopping rules is significantly less for Type A than for Type B expansions. And there is still another reason why expansions of Type A are the "better" ones: From (5.47), we obtain the relation

$$b^{\rho_{k+1}} e_{i+1}^{(k)} = e_i^{(k)} + (b^{\rho_{k+1}} - 1) e_i^{(k+1)} ,$$

from which it follows that, for Type A expansions, we have for all $k \geq k_0$,

$$\frac{e_i^{(k)}}{e_{i+1}^{(k)}} \geq b^{\rho_{k+1}} ,$$

whereas, for Type B ones,

$$\frac{e_i^{(k)}}{e_{i+1}^{(k)}} \leq b^{\rho_{k+1}} .$$

This means that Type A expansions lead to a *better rate of convergence* in each column of the extrapolated tableau than the Type B expansions.

If extrapolation is considered as a method for the approximation of functions, i.e., if the initial sequence $\{\sigma_n\} = \{\sigma_n(x)\}$ depends on an additional variable x, then normality in our sense can also be used to characterize situations, where the resulting error *functions* $e_i^{(k)}(x)$ have a predictionable monotonicity behaviour, since, just like in Theorem 5.5 the behaviour of $\text{sign}(e_i^{(k)})$ was reduced to that of $\text{sign}(c_{k+1})$, one can also find a dependence of the behaviour of $e_i^{(k)}(x)$ as a function of x on that of $c_{k+1}(x)$:

Lemma 5.8 (G.WALZ [4]): *Let $\{\sigma_n(x)\}$ be a sequence of functions, defined on some interval I and possessing a k_0- normal asymptotic expansion. Furthermore, denote for each $x \in I$ by $e_i^{(k)}(x)$ the error of the extrapolated values $y_i^{(k)}(x)$.*

If there is an interval $I_1 \subset I$, such that

$$\left|c_{k+1}^{(k)}(x_2) - c_{k+1}^{(k)}(x_1)\right| > \left|c_\mu^{(k)}(x_2) - c_\mu^{(k)}(x_1)\right|$$

for all $k_0 \leq k < m$, $\mu > k+1$ and $x_1 < x_2$, $x_1, x_2 \in I_1$, then

$$\text{sign}\,(e_i^{(k)}(x_2) - e_i^{(k)}(x_1)) = (-1)^k \cdot \text{sign}\,(c_{k+1}(x_2) - c_{k+1}(x_1))$$

for all $i \in \mathbb{N}$.

The proof of Lemma 5.8 is of course in complete analogy to that of Theorem 5.5 and therefore omitted here.

It is evident how the statement of the lemma can be used to predict the monotonicity behaviour of the error function $e_i^{(k)}(x)$: Let $I_2 \subset I_1$ be an interval, in which $c_{k+1}(x)$ is a monotone fuction of x, without loss of generality an increasing function. Then $e_i^{(k)}$ is also monotone on I_2, it is increasing, if k is even, and decreasing, if k is odd. In particular it should be noted that, for fixed k, all functions $e_i^{(k)}$ show the same monotonicity behaviour (independent of i).

Some numerical examples for this effect are contained in G.WALZ [4]. Also, additional information of the approximation of functions by extrapolation methods can be found in Section 3.3 or S.FILIPPI [1], H.RUTISHAUSER [1], and G.WALZ [1,3].

5.2.3 Examples. The various situations and cases of normality are illustrated best by looking at the several methods for the computation of π.

The first approach, initiated in ARCHIMEDES [1] and later worked out in CH. HUYGENS [1] and J.F.SAIGEY [1] (see the Historical Notes), is the following one: Denoting again by A_n the area of the $n-$ sided regular polygon, which is inscribed into the unit circle, we already had seen in Subsection 3.1.1 that the values

$$y_i^{(0)} := A_{2^{i+2}}$$

can be computed by means of the recursion formula

$$\begin{aligned} y_0^{(0)} &:= 2\,, \\ y_{i+1}^{(0)} &:= y_i^{(0)} \cdot \frac{\sqrt{2}}{\sqrt{1+\sqrt{1-(y_i^{(0)}/2^i)^2}}} \qquad i = 0, 1, \dots\,. \end{aligned} \tag{5.52}$$

Moreover, it was already shown there that A_{2n} has the explicit representation

$$A_{2n} \;=\; n \cdot \sin\Big(\frac{\pi}{n}\Big)$$

and therefore the sequence $\{A_n\}$ possesses the asymptotic expansion of arbitrary order

$$A_{2n} \;=\; \pi + \sum_{\mu=1} \frac{c_\mu}{n^{2\mu}} \tag{5.53a}$$

with

$$c_\mu \;=\; (-1)^\mu \cdot \frac{\pi^{2\mu+1}}{(2\mu+1)!} \tag{5.53b}$$

for all $\mu \in I\!N$, which is now easily recognized to be normal of Type B. (To be somewhat overprecise: The sequence $\{-A_n\}$ is $0-$ normal of Type B in the sense of Definition 5.3).

Thanks to (5.52), we can compute all values $y_i^{(0)} = A_{2^{i+2}}$ and therefore can apply the standard extrapolation process (4.31) (resp. (5.5)), using $b = 2$.

This method, which we denote for historical reasons as *(first) Saigey method* or (*Saigey I* for short) thus produces sequences of approximations $\{y_i^{(k)}\}$ to the number π, and in view of Theorem 5.5 we expect *all* errors $e_i^{(k)} := y_i^{(k)} - \pi$ to be negative.

This is illustrated quite nicely in Table 5.3a, which shows us these errors $e_i^{(k)}$ for $k = 0, \ldots, 4$ and $i = 0, \ldots, 6 - k$.

−.313 e(00)				
	−.245 e(-2)			
−.801 e(-1)		−.226 e(-5)		
	−.155 e(-3)		−.304 e(-9)	
−.201 e(-1)		−.356 e(-7)		−.667 e(-14)
	−.972 e(-5)		−.119 e(-11)	
−.504 e(-2)		−.558 e(-9)		−.654 e(-17)
	−.608 e(-6)		−.467 e(-14)	
−.126 e(-2)		−.872 e(-11)		−.639 e(-20)
	−.380 e(-7)		−.182 e(-16)	
−.315 e(-3)		−.136 e(-12)		
	−.237 e(-8)			
−.789 e(-4)				

Table 5.3a. Approximating π using "Saigey I": Errors $e_i^{(k)}$

Archimedes also proposed another method for computing π numerically, namely to consider the areas U_n of the regular $n-$ sided polygons, which are *circumscribed* to the unit circle. By elementary geometrical considerations it is clear that, for $n \geq 3$,

$$U_n = n \cdot \tan\left(\frac{\pi}{n}\right) \tag{5.54}$$

and, moreover, that for each pair of natural numbers (n_1, n_2) (with $n_2 \geq 3$) the inclusion

$$A_{n_1} < \pi < U_{n_2} \tag{5.55}$$

holds (cf. (5.3)).

Note that for the sequence $\{U_n\}$ the situation is completely analoguos to that of $\{A_n\}$: Also here, the subsequence which belongs to the indices 2^i, i.e. the values

$$u_i^{(0)} := U_{2^{i+2}}$$

can be computed through

$$\begin{aligned} u_0^{(0)} &:= 4\,, \\ u_{i+1}^{(0)} &:= u_i^{(0)} \cdot \frac{2}{\sqrt{1+\sqrt{1+(u_i^{(0)}/2^i)^2}}} \qquad i = 0, 1, \dots\,, \end{aligned} \tag{5.56}$$

(see Problem P 3.8), and, using Taylor expansion, it is easily recognized that the sequence $\{U_n\}$ possesses the asymptotic expansion of arbitrary order

$$U_n = \pi + \sum_{\mu=1} \frac{d_\mu}{n^{2\mu}} \tag{5.57a}$$

with

$$d_\mu = \frac{(-1)^\mu \cdot \pi^{2\mu+1} \cdot 2^{2\mu+2}(2^{2\mu+2}-1) \cdot \beta_{2\mu+2}}{(2\mu+2)!}\,, \tag{5.57b}$$

$\{\beta_{2\mu+2}\}$ denoting the Bernoulli numbers, which were studied extensively in Section 3.2.

Therefore it is justified to apply the standard extrapolation process (4.31) (resp. (5.5)) with $b = 2$ to the sequence $\{u_i^{(0)}\}$, which produces the extrapolated sequences, say, $\{u_i^{(k)}\}$; this method is called *(second) Saigey method* or *Saigey II* for short.

Now, remember for a moment Relation (5.55), which was naturally of outstanding interest to Archimedes and the other ancient mathematicians and geometers; therefore, it was hoped for some time that an inclusion of this type is also true for the extrapolated sequences, i.e., that $y_i^{(k)} < \pi < u_j^{(k)}$ holds for all $k \in \mathbb{N}$ and each pair (i, j).

But now we see that this cannot be true, since, owing to the sign behaviour of the Bernoulli numbers, the asymptotic expansion (5.57) of the sequence $\{U_n\}$ is normal of Type A, and so, following Theorem 5.5, the errors $\tilde{e}_i^{(k)} := u_i^{(k)} - \pi$ alternate in sign, as k increases; see Table 5.3b for a numerical illustration of this effect.

.858 e(00)				
	−.556 e(-1)			
.172 e(00)		.898 e(-3)		
	−.270 e(-2)		−.352 e(-5)	
.410 e(-1)		.106 e(-4)		.344 e(-8)
	−.159 e(-3)		−.103 e(-7)	
.101 e(-1)		.155 e(-6)		.252 e(-11)
	−.978 e(-5)		−.378 e(-10)	
.253 e(-2)		.238 e(-8)		.231 e(-14)
	−.609 e(-6)		−.146 e(-12)	
.631 e(-3)		.371 e(-10)		
	−.380 e(-7)			
.158 e(-3)				

Table 5.3b. Approximating π using "Saigey II": Errors $\tilde{e}_i^{(k)}$

So, at first sight this seems to be a negative result. But, quite to the contrary, a closer look onto this matter convinces us that the result is indeed very positive, since already the $u_i^{(k)}$'s *alone* include the true value in a number of ways, since, again according to Theorem 5.5, the relations

$$u_i^{(2\kappa_1)} > \pi > u_j^{(2\kappa_2+1)} \tag{5.58}$$

hold for all $\kappa_1, \kappa_2 \in \mathbb{N}_0$ and each pair of indices (i, j).

Let us study now another, not so well-known method for computing π, which is due to K.KOMMERELL [1]; for details on the historical development, please see G.WALZ [11] or have a look into the Historical Notes at the end of this book.

Kommerell's idea can be described as follows: Starting with a regular $n-$ sided polygon P_n of fixed circumference 1, one constructs the circle C_n which is circumscribed to P_n, and the circle I_n, which is inscribed into P_n. Now one computes iteratively polygons with $2n$, $4n$, $8n, \dots$ vertices such that – in contrast to Archimedes' resp. Saigey's method – the circumference 1 remains fixed. This implies of course that the corresponding radii r_n of C_n and ρ_n of I_n converge to that of a circle with circumference 1, i.e. we have

$$\lim_{n\to\infty} \rho_n = \lim_{n\to\infty} r_n = \frac{1}{2\pi}, \tag{5.59}$$

from which the number π can be computed. Kommerell did not provide explicit formulae for the calculation of ρ_n and r_n, but it is easy to see that, for $n \geq 2$,

$$\begin{aligned} r_n &= \frac{1}{2n \cdot \sin(\pi/n)} \quad \text{and} \\ \rho_n &= r_n \cdot \cos(\pi/n) = \frac{1}{2n \cdot \tan(\pi/n)}. \end{aligned} \tag{5.60}$$

The series expansions of the functions $1/\sin(z)$ and $1/\tan(z)$ (see e.g. M.ABRAMOWITZ & I.STEGUN [1]) imply at once that both sequences under consideration possess a logarithmic asymptotic expansion of arbitrary order with limit $1/(2\pi)$; we have

$$\begin{aligned} r_n &= \frac{1}{2\pi} + \sum_{\mu=1} \frac{c_\mu^r}{n^{2\mu}}, \quad \text{and} \\ \rho_n &= \frac{1}{2\pi} + \sum_{\mu=1} \frac{c_\mu^\rho}{n^{2\mu}} \end{aligned} \tag{5.61a}$$

with the coefficients

$$\begin{aligned} c_\mu^r &= \frac{(-1)^{\mu+1}\cdot\pi^{2\mu-1}\cdot(2^{2\mu-1}-1)\cdot\beta_{2\mu}}{(2\mu)!}\,, \qquad \text{and} \\ c_\mu^\rho &= \frac{(-1)^{\mu}\cdot\pi^{2\mu-1}\cdot 2^{2\mu}\cdot\beta_{2\mu}}{(2\mu)!}\,. \end{aligned} \tag{5.61b}$$

For practical use it is of great importance that again those subsequences, which are defined by repeated index-doubling, can be computed *recursively*, since, for $n \geq 2$, the relations

$$\begin{aligned} \rho_{2n} &= \frac{r_n+\rho_n}{2} \qquad \text{and} \\ r_{2n} &= \sqrt{r_n\cdot\rho_{2n}} \end{aligned} \tag{5.62}$$

hold. Kommerell himself proved these relations by means of geometrical considerations, but they can also be derived from (5.60): Using the half-angle formulae for $\tan(z)$ and $\sin(z)$ (with $z = \pi/n$), we find that

$$\begin{aligned} \rho_{2n} &= \frac{1}{4n\cdot\tan(\pi/2n)} \\ &= \frac{1}{4n}\cdot\frac{1+\cos(\pi/n)}{\sin(\pi/n)} = \frac{r_n+\rho_n}{2}\,, \end{aligned}$$

and, similarly,

$$\begin{aligned} r_n\cdot\rho_{2n} &= r_{2n}\cdot\frac{\cos(\pi/2n)}{2n\cdot\sin(\pi/n)} \\ &= r_{2n}\cdot\frac{\cos(\pi/2n)}{4n\cdot\sin(\pi/2n)\cdot\cos(\pi/2n)} = r_{2n}^2\,, \end{aligned}$$

which proves (5.62).

Now note that we are once again in the nice situation that the sequences under consideration include the true value $1/2\pi$, since we have

$$\rho_2 < \rho_3 < \cdots < \rho_n < \cdots < \frac{1}{2\pi} < \cdots < r_n < \cdots < r_3 < r_2 .$$

But Kommerell was even not satisfied with this approximation, and he presented – as a new result – a convergence accelerating method; by very subtle calculations he proved that the sequences $\{R_n\}$ and $\{S_n\}$, defined through

$$\begin{aligned} R_n &:= \frac{4\,r_{2n} - r_n}{3} \quad \text{and} \\ S_n &:= \frac{4\,\rho_{2n} - \rho_n}{3}\,, \end{aligned} \tag{5.63}$$

provide better approximations to the limit than the old ones, and moreover, that still the inclusion

$$R_{n_1} < \frac{1}{2\pi} < S_{n_2} \tag{5.64}$$

is true for each pair of natural numbers (n_1, n_2).

Obviously, formulae (5.63) are nothing else but the first step of the standard extrapolation process (4.31) with $b = 2$, and so the situation is now different from that in Saigey's approach, since still the pairs of (first) extrapolates includes the true value. Naturally, Kommerell didn't know anything about the asymptotic expansion (5.61), nor could he prove inclusion statements for higher order extrapolations. But we can, and summarizing our above-made considerations we obtain the following result:

The sequences $\{r_n\}$ and $\{-\rho_n\}$, with r_n and ρ_n defined through (5.60), are both $0-$ normal of Type A; therefore, denoting the extrapolates of r_n by $y_i^{r\,(k)}$, and those of ρ_n by $y_i^{\rho\,(k)}$, the sign behaviour of the corresponding errors

$$e_i^{r\,(k)} := y_i^{r\,(k)} - \frac{1}{2\pi} \quad \text{and} \quad e_i^{\rho\,(k)} := y_i^{\rho\,(k)} - \frac{1}{2\pi}$$

can be characterized as follows: We have

$$\begin{aligned} \operatorname{sign}\bigl(e_i^{r\,(k)}\bigr) &= (-1)^k \qquad \text{and} \\ \operatorname{sign}\bigl(e_i^{\rho\,(k)}\bigr) &= (-1)^{k+1} \end{aligned} \tag{5.65}$$

for all i and k. This provides at once a lot of possibilities for an inclusion of the true value; for example, we have

$$y_i^{\rho\,(k)} < \frac{1}{2\pi} < y_i^{r\,(k)},$$

if k is even, and

$$y_i^{r\,(k)} < \frac{1}{2\pi} < y_i^{\rho\,(k)},$$

if k is odd; moreover, two subsequent columns of *each* of the both schemes always establish an inclusion.

A numerical illustration of these statements is given in Table 5.4, whose entries are the just defined errors $e_i^{r\,(k)}$ and $e_i^{\rho\,(k)}$, resp., for $k = 0, \ldots, 4$ and $i = 2, \ldots, 8-k$.

.176 e(-1)				
	−.320 e(-3)			
.417 e(-2)		.130 e(-5)		
	−.188 e(-4)		−.129 e(-8)	
.103 e(-2)		.191 e(-7)		.315 e(-12)
	−.116 e(-5)		−.471 e(-11)	
.256 e(-3)		.294 e(-9)		.288 e(-15)
	−.720 e(-7)		−.181 e(-13)	
.639 e(-4)		.457 e(-11)		.277 e(-18)
	−.449 e(-8)		−.705 e(-16)	
.160 e(-4)		.714 e(-13)		
	−.281 e(-9)			
.399 e(-5)				

Table 5.4a. Approximating $1/(2\pi)$ using Kommerell's method: Errors $e_i^{r\,(k)}$

−.342 e(-1)				
	.363 e(-3)			
−.827 e(-2)		−.134 e(-5)		
	.214 e(-4)		.130 e(-8)	
−.205 e(-2)		−.197 e(-7)		−.316 e(-12)
	.132 e(-5)		.475 e(-11)	
−.512 e(-3)		−.303 e(-9)		−.289 e(-15)
	.822 e(-7)		.183 e(-13)	
−.128 e(-3)		−.472 e(-11)		−.278 e(-18)
	.514 e(-8)		.711 e(-16)	
−.320 e(-4)		−.737 e(-13)		
	.321 e(-9)			
−.799 e(-5)				

Table 5.4b. Approximating $1/(2\pi)$ using Kommerell's method: Errors $e_i^{\rho\,(k)}$

So far we have considered some examples on the sign behaviour of normal asymptotic expansions and their extrapolates, thus illustrating the statements of Theorem 5.5; now, let us come to a numerical example for the second main topic of this section, the stopping rules.

Surely, the rule for Type A expansions (Theorem 5.6) is to a large extend self-explaining, and its importance for the practice is clear, so let us turn immediately to Type B expansions. Here, we want to illustrate the statement of Theorem 5.7, i.e., the stopping rule governed by Relation (5.46); our example is the above-discussed method Saigey I, which is based on the existence and Type B normality of the asymptotic expansion

$$n \cdot \sin\left(\frac{\pi}{n}\right) = \pi + \sum_{\mu=1} \frac{c_\mu}{n^{2\mu}}$$

with the coefficients

$$c_\mu = (-1)^\mu \cdot \frac{\pi^{2\mu+1}}{(2\mu+1)!}\,,$$

see (5.53).

.313 e(00)				
	.245 e(-2)			
	.333 e(-2)			
	.338 e(-2)			
.801 e(-1)		.226 e(-5)		
		.305 e(-5)		
		.309 e(-5)		
	.155 e(-3)		.304 e(-9)	
	.208 e(-3)		.408 e(-9)	
	.210 e(-3)		.414 e(-9)	
.201 e(-1)		.356 e(-7)		.667 e(-14)
		.476 e(-7)		.894 e(-14)
		.483 e(-7)		.906 e(-14)
	.972 e(-5)		.119 e(-11)	
	.130 e(-4)		.159 e(-11)	
	.131 e(-4)		.161 e(-11)	
.504 e(-2)		.558 e(-9)		.654 e(-17)
		.744 e(-9)		.873 e(-17)
		.754 e(-9)		.884 e(-17)
	.608 e(-6)		.467 e(-14)	
	.811 e(-6)		.623 e(-14)	
	.821 e(-6)		.631 e(-14)	
.126 e(-2)		.872 e(-11)		.639 e(-20)
		.116 e(-10)		.852 e(-20)
		.118 e(-10)		.864 e(-20)
	.380 e(-7)		.182 e(-16)	
	.507 e(-7)		.243 e(-16)	
	.513 e(-7)		.246 e(-16)	
.315 e(-3)		.136 e(-12)		
		.182 e(-12)		
		.184 e(-12)		
	.237 e(-8)			
	.317 e(-8)			
	.321 e(-8)			
.789 e(-4)				

Table 5.5. Test of stopping rule (5.46) (see text)

In order to use the right hand side of (5.46) as a stopping criterion, we first have to choose suitable values for the $\alpha_i^{(k)}$'s; according to (5.51), we can put in this example

$$\alpha_i^{(k)} := 4^{k+1}\left(1 - \frac{\pi^2}{(2k+4)(2k+5)} \cdot \frac{1-4^{k+1}}{4^k - 4^{k+1}} \cdot 4^{-i}\right) \qquad (5.66)$$

for all i and k.

This turns out to be a suitable choice here and gives nice results, but it might still suffer from the fact that we make use of the number π in an error estimate for an approximation of precisely this number! To avoid this, we can replace π^2 in (5.66) by the rough estimate 10.

The output of our corresponding numerical illustration is shown in Table 5.5; the entries at each place of this tableau are the absolute values of the following numbers: True error (top), estimate on the right hand side of (5.46) with $\alpha_i^{(k)}$ as in (5.66) (middle), and the same with π^2 replaced by 10. Obviously, the estimation of the true errors is quite good for each of the both choices.

We close this subsection on examples for normal asymptotic expansions by remarking once more that normality is by far not restricted to the approximation of π, but it is also a natural property of asymptotic expansions which occur in connection with the approximation of functions (cf. Section 3.3) or with numerical differentiation (cf. Section 3.5); e.g., we conclude from Theorem 3.55 that for each fixed real number $x > 1$ the sequence $\{y_n(1,x)\}$, defined through

$$y_n(1,x) := n \cdot (x^{1/n} - 1),$$

possesses the logarithmic asymptotic expansion

$$y_n(1,x) = \log(x) + \sum_{\mu=1} \frac{c_\mu(1,x)}{n^\mu} \qquad (5.67a)$$

with the coefficients

$$c_\mu(1,x) \;=\; \frac{\log(x)^{\mu+1}}{(\mu+1)!} \qquad (5.67b),$$

which is normal of Type A; for numerical results on this and some more examples, see G.WALZ [4].

5.3 Miscellanea

In addition to the results mentioned so far, there are also some more considerations concerning the error of extrapolation processes in the existing literature. For example, a quite general approach to the subject was taken in C.BREZINSKI [3]. Moreover, especially within the context of ODE-solvers, error questions always played a very prominent role; for more information on this, the reader is kindly referred to R.BULIRSCH & J.STOER [2], P.DEUFLHARD [1], L.F.SHAMPINE & L.S.BACA [1,2], or J.STOER [2], to quote only a few outstanding publications on the subject.

In the rest of this section we concentrate on two other topics, which are of interest in connection with extrapolation methods, namely the computation of bounds for the propagated rounding error, and the preservation of monotonicity (without having any information on an underlying asymptotic expansion) by the extrapolation process.

5.3.1 Propagation of Rounding Errors. In this subsection, we are interested in the numerical stability of the logarithmic extrapolation processes; first, let us consider the standard logarithmic extrapolation process (4.31) resp. (5.5), i.e., our logarithmic asymptotic expansion is assumed to be of the general form

$$\sigma_n \;=\; c_0 \;+\; \sum_{\mu=1}^{m} \frac{c_\mu}{n^{\rho_\mu}} \;+\; o(n^{-Re\rho_m}), \qquad (5.68)$$

and with $y_i^{(0)} := \sigma_{n_0 \cdot b^i}$, all $y_i^{(k)}$ are computed through

$$y_i^{(k)} = y_{i+1}^{(k-1)} + \frac{y_{i+1}^{(k-1)} - y_i^{(k-1)}}{b^{\rho_k} - 1} . \tag{5.69}$$

We want to compute a bound for the propagated rounding error in the k-th column of this tableau. Building on earlier work, concerning the Romberg case (i.e. $b = 2$ and $\rho_k = 2k$ for all k), by F.L.BAUER, H.RUTISHAUSER & E.STIEFEL [1], the following result was proved by J.A.Shanks:

Theorem 5.9 (see J.A.SHANKS [1]): *For all $0 \leq k \leq m$, let $R^{(k)}$ denote the bound of the propagated rounding error in the k-th column of the $y_i^{(k)}$– tableau. Then*

$$R^{(k)} = R^{(0)} \cdot \prod_{\mu=1}^{k} \frac{b^{\rho_\mu} + 1}{b^{\rho_\mu} - 1} . \tag{5.70}$$

Furthermore, if, in the case $m = \infty$, there is a constant $\eta > 0$, such that

$$\rho_{\mu+1} - \rho_\mu \geq \eta \quad \textit{for all } \mu , \tag{5.71}$$

then

$$\lim_{k \to \infty} \prod_{\mu=1}^{k} \frac{b^{\rho_\mu} + 1}{b^{\rho_\mu} - 1} \tag{5.72}$$

exists, which implies numerical stability in the (infinite) tableau.

Proof. Due to (5.69), the $R^{(k)}$'s satisfy

$$R^{(k)} = R^{(k-1)} \cdot \left(1 + \frac{2}{b^{\rho_k} - 1}\right) = R^{(k-1)} \cdot \frac{b^{\rho_k} + 1}{b^{\rho_k} - 1} ,$$

from which (5.70) already follows.

To prove the second statement, we note that the infinite product in (5.72) exists, if and only if the series

$$\sum_{\mu=1}^{\infty} \frac{2}{b^{\rho_\mu} - 1}$$

converges (see e.g. K.KNOPP [1]). But, due to the quotient criterion, this is indeed the case, if (5.71) is satisfied, and so the theorem is proved. □

For example, numerical values of the infinite product in (5.72) for $b = 2$ and $\rho_\mu = \mu$ resp. $\rho_\mu = 2\mu$ are

$$\prod_{\mu=1}^{\infty} \frac{2^\mu + 1}{2^\mu - 1} = 8.2559879\ldots$$

and

$$\prod_{\mu=1}^{\infty} \frac{4^\mu + 1}{4^\mu - 1} = 1.96926035\ldots$$

Moreover, it is clear now that numerical stability increases with η; this corresponds to our remark following Definition 3.40 concerning the influence of the modulus η on the numerical stability of the extrapolation process.

Now, let us have a look at the modified extrapolation process (4.45) resp. (5.6); here we are forced to assume that the asymptotic expansion of the initial sequence is a special one, and, moreover, we consider it to be of the form

$$\sigma_n = c_0 + \sum_{\mu=1} \frac{c_\mu}{n^{2\mu}} . \tag{5.73}$$

As index sequence, we use a quite extremal one, namely the Harmonic sequence $n_i = (i+1)$; although here a very bad numerical behaviour is expected, at least the roundoff errors in the top diagonal sequence $y_0^{(k)}$ can be estimated. The extrapolation process reads in this case

$$y_i^{(k)} = y_{i+1}^{(k-1)} + \frac{(i+1)^2}{(i+k+1)^2 - (i+1)^2} \cdot \left(y_{i+1}^{(k-1)} - y_i^{(k-1)} \right) , \tag{5.74}$$

and since the factors $(i+1)^2/((i+k+1)^2 - (i+1)^2)$ depend on i, it is convenient to introduce also for the $R^{(k)}$'s a second index and to consider $R_i^{(k)}$, the propagated roundoff error in $y_i^{(k)}$, where of course $R_i^{(0)} \equiv R^{(0)}$ for all i.

Theorem 5.10 (D.P.LAURIE [1]): *For all* $k \geq 0$, *the bounds* $R_0^{(k)}$ *can be estimated by*

$$R_0^{(k)} < \frac{C \cdot M^k}{\sqrt{k+\frac{1}{2}}} \tag{5.75}$$

with the constants

$$C = \frac{\exp(2)}{4\sqrt{\pi}} = 1.0422\ldots, \quad \textit{and} \tag{5.76a}$$

$$M = \exp(2)/3 = 2.4630\ldots. \tag{5.76b}$$

Proof. From (5.74) we recognize that the $R_i^{(k)}$'s satisfy the relation

$$R_i^{(k)} = R_{i+1}^{(k-1)} + \frac{(i+1)^2}{(i+k+1)^2-(i+1)^2} \cdot \left(R_{i+1}^{(k-1)} + R_i^{(k-1)}\right), \tag{5.77}$$

from which we conclude by induction that $R_i^{(k)} \leq R_{i+1}^{(k)}$ for all i, and so, using (5.77) again,

$$R_i^{(k)} \leq R_{i+1}^{(k-1)} \cdot \left(1+\frac{2(i+1)^2}{(i+k+1)^2-(i+1)^2}\right). \tag{5.78}$$

Applying Relation (5.78) several times, we obtain the following chain of inequalities:

$$\begin{aligned} R_0^{(k)} &\leq R^{(0)} \cdot \prod_{n=1}^{k}\left(1+\frac{2(k-n+1)^2}{(k+1)^2-(k-n+1)^2}\right) \\ &= R^{(0)} \cdot \prod_{n=1}^{k}\left(1+\frac{2n^2}{(k+1)^2-n^2}\right) \\ &= R^{(0)} \cdot \prod_{n=1}^{k}\frac{(k+1)^2+n^2}{(k+n+1)(k-n+1)} \end{aligned}$$

$$= R^{(0)} \cdot \frac{k+1}{(2k+1)!} \cdot \prod_{n=1}^{k} \left((k+1)^2 + n^2\right)$$

$$\leq R^{(0)} \cdot \frac{k+1}{(2k+1)!} \cdot \left[\frac{1}{k} \cdot \sum_{n=1}^{k} \left((k+1)^2 + n^2\right)\right]^k .$$

Now we use the relations

$$\sum_{n=1}^{k} n^2 = \frac{k(k+1)(2k+1)}{6}$$

and

$$\frac{2k+1}{6(k+1)} < \frac{1}{3},$$

valid for all k, and obtain that

$$R_0^{(k)} = R^{(0)} \cdot \frac{(k+1)^{2k+1}}{(2k+1)!} \cdot \left[1 + \frac{2k+1}{6(k+1)}\right]^k$$

$$< R^{(0)} \cdot \frac{(k+1)^{2k+1}}{(2k+1)!} \cdot \left(\frac{4}{3}\right)^k .$$

Now, the statement of Theorem 5.10 follows by means of Stirling's inequality. □

5.3.2 Monotonicity Preserving Extrapolation. Let us now consider the following question, which was originally investigated in a little series of papers by G. OPFER [1,2,3]: Given a sequence $\{x_n\}$ of real numbers, converging monotonically to some limit ξ; what conditions can be established such that the convergence of the "extrapolated sequence"

$$y_\nu := \frac{x_{\nu+1} - \gamma_\nu \cdot x_\nu}{1 - \gamma_\nu} \tag{5.79}$$

(γ_ν prescribed real numbers) is also monotone, either in the same direction as $\{x_\nu\}$, or in the opposite one. Clearly, the situation that $\{y_\nu\}$ converges from the opposite side to ξ is more desirable, since then we have an inclusion of this limit by each pair (x_{ν_1}, y_{ν_2}); therefore we shall concentrate on this case (i.e. inversion of the monotonicity direction). Moreover, without serious restriction of generality we assume $\{x_\nu\}$ to be decreasing, i.e.

$$x_{\nu+1} < x_\nu \quad \text{for all } \nu . \tag{5.80}$$

Note that Process (5.79) establishes a very general approach, covering for example (one step of) the geometric extrapolation (in particular Aitken's Δ^2- process, see Subsection 4.1.1), as well as the several logarithmic extrapolation methods we have investigated; on the other hand, the result we are going to present does not regard the possible existence of an asymptotic expansion for the underlying sequence $\{x_\nu\}$ (and this is why we treat it under "miscellanea"; corresponding results on asymptotic expansions were considered at length in the previous subsections).

In the sequel, we shall repeatedly make use of the ratios

$$c_\nu := \frac{x_{\nu+2} - x_{\nu+1}}{x_{\nu+1} - x_\nu} \tag{5.81}$$

which are positive for all ν, due to (5.80). Now we can state the following main result:

Theorem 5.11 (G.Opfer [1,3]): *Consider a sequence $\{x_\nu\}$ of real numbers, satisfying (5.80) and converging to some limit ξ, and define the sequence $\{y_\nu\}$ through (5.79), with a given set $\{\gamma_\nu\}$, not converging to 1, and such that*

$$\gamma_{\nu+1} \le \gamma_\nu \quad \textit{and} \quad \gamma_\nu \ne 1 \quad \textit{for all } \nu . \tag{5.82}$$

Then $\{y_\nu\}$ is an increasing sequence with limit ξ, if and only if

$$0 < \gamma_1 < 1 \quad \textit{and} \quad c_\nu \le \frac{\gamma_\nu(1-\gamma_{\nu+1})}{1-\gamma_\nu} \quad \textit{for all } \nu . \tag{5.83}$$

Moreover, sufficient for this kind of convergence is that

$$c_\nu \le \gamma_\nu < 1 \quad \textit{for all } \nu . \tag{5.84}$$

Proof. We begin by stating two easy observations: First, due to (5.82) and to the fact that $c_\nu > 0$ for all ν, we see that (5.83) is equivalent to

$$0 < \gamma_\nu < 1 \quad \text{and} \quad c_\nu \le \frac{\gamma_\nu(1-\gamma_{\nu+1})}{1-\gamma_\nu} \quad \text{for all } \nu\,. \tag{5.83'}$$

Secondly, it is obvious that Relation (5.84) implies (5.83'). So we are left to prove the equivalence of (5.83') with the claimed monotonicity behaviour of $\{y_\nu\}$.

If $\{y_\nu\}$ is increasing, then by definition

$$\frac{x_{\nu+1} - \gamma_\nu \cdot x_\nu}{1-\gamma_\nu} \le \frac{x_{\nu+2} - \gamma_{\nu+1} \cdot x_{\nu+1}}{1-\gamma_{\nu+1}} \tag{5.85}$$

holds for all ν.

We now distinguish two cases (remember that $\gamma_\nu \neq 1$):
(i) $1-\gamma_\nu > 0$ for all ν. We then find successively that (5.85) is equivalent to

$$\frac{1-\gamma_{\nu+1}}{1-\gamma_\nu} \cdot \left(x_{\nu+1} - \gamma_\nu \cdot x_\nu\right) \le \left(x_{\nu+2} - \gamma_{\nu+1} \cdot x_{\nu+1}\right), \quad \text{to}$$

$$\frac{1-\gamma_{\nu+1}}{1-\gamma_\nu} \cdot \frac{\gamma_\nu \cdot x_\nu - x_{\nu+1}}{x_\nu - x_{\nu+1}} \ge \frac{\gamma_{\nu+1} \cdot x_{\nu+1} - x_{\nu+2}}{x_\nu - x_{\nu+1}}, \quad \text{and to}$$

$$\gamma_\nu \cdot \frac{1-\gamma_{\nu+1}}{1-\gamma_\nu} \ge c_\nu\,,$$

i.e., the second term in (5.83'); since $c_\nu > 0$, this implies in turn that $0 < \gamma_\nu$ for all ν, and thus (5.83') is proved in this case.

(ii) There exists an index m with $1-\gamma_m < 0$. Then

$$y_m - \xi = \frac{\gamma_m(x_m - \xi) - (x_{m+1} - \xi)}{\gamma_m - 1} > 0\,, \tag{5.86}$$

contradicting the fact that $\{y_\nu\}$ is increasing and converging to ξ. Therefore, is it impossible that such an m exists, and the necessity of (5.83) is proved.

Finally, the sufficiency of this condition follows by inverting the proof under (i). □

The statement of Theorem 5.11 applies to one single step of the geometric resp. logarithmic extrapolation process. To prove some corresponding results on a multiple extrapolation step, that is, to express the monotonicity behaviour of the kth extrapolates through that of the initial sequence, more sophisticated methods are needed; this was done quite successfully in G.OPFER [2], see also G.OPFER [3].

As an example for the process (5.79) we take an arbitray, say the first step of our (standard) logarithmic extrapolation process (4.31); here,

$$\gamma_\nu \equiv \gamma = b^{-\rho_1} \quad \text{for all } \nu\,, \tag{5.87}$$

hence $0 < \gamma < 1$. From Theorem 5.11 we learn that the process inverts the monotonicity direction, if $c_\nu \leq b^{-\rho_1}$ for all ν; in the Romberg case, this just means $c_\nu \leq \frac{1}{4}$.

Similar remarks apply to the modified logarithmic extrapolation (4.45), where

$$0 < \gamma_\nu = \left(\frac{n_\nu}{n_{\nu+k}}\right)^\eta < 1 \tag{5.88}$$

for all ν, as well as to the geometric extrapolation process (4.2).

Another very interesting example is given by what we called the approximative geometric extrapolation process, and in particular by Aitken's Δ^2-process, which is a special case of it, see Lemma 4.2 and the subsequent remarks. Here, in the notation of Theorem 5.11, one puts $\gamma_\nu := c_\nu$ for all ν, and the theorem implies that the process inverts the monotonicity direction of a decreasing sequence $\{x_\nu\}$, if $\{c_\nu\}$ is decreasing with $c_1 < 1$.

In an a little bit different setting (so-to-say, from an asymptotic point of view), this can also be read off from Theorem 5.2.

5.4 Problems

P 5.1. State (and prove) a result analogue to that of Part a) in Theorem 5.1 for the geometric extrapolation process (4.2).
Hint: Put $\lambda_\mu := b^{-\rho_\mu}$ and try to re-formulate the geometric extrapolation as a (standard) logarithmic extrapolation.

P 5.2. Prove Relation (5.13a) in Theorem 5.1.
Hint: Consider first the specific logarithmic asymptotic expansion

$$\sigma_n \;:=\; n^{-\eta j}$$

with $j \in I\!N$ fixed, and then use the linearity of the extrapolation process (5.6) (cf. J.W.Schmidt [1]).

P 5.3 (cf. R.Bulirsch & J.Stoer [3]). If we replace the definition of α_k in (5.8) by simply setting $\alpha_k = 1$ for all k, then (5.7) can be interpreted as the result of the (standard) extrapolation process, with $b = 2$, applied to the sequence

$$u_n \;:=\; 2\,\sigma_{2n} - \sigma_n \;.$$

Now assume for simplicity that σ_n possesses an asymptotic expansion of the form

$$\sigma_n \;=\; c_0 + \sum_{\mu=1} \frac{c_\mu}{n^{\eta\mu}}$$

with $c_\mu \neq 0$ for all μ.

Prove that u_n then possesses the asymptotic expansion

$$u_n \;=\; c_0 + \sum_{\mu=1} \frac{d_\mu}{n^{\eta\mu}}$$

with

$$\operatorname{sign} c_\mu = -\operatorname{sign} d_\mu \quad \text{for all } \mu ,$$

and

$$\lim_{\mu\to\infty} \frac{c_\mu}{d_\mu} = -1 .$$

P 5.4. a) Prove Relations (5.61), i.e., show that the sequences $\{r_n\}$ and $\{\rho_n\}$, defined in (5.60), possess a logarithmic asymptotic expansion of arbitrary order with limit $1/(2\pi)$.
b) Prove explicitly that the sequences $\{r_n\}$ and $\{-\rho_n\}$ are $0-$ normal of type A.

P 5.5. State (and prove) a normality result on the sequences $\{y_n(\alpha, x)\}$, defined in Theorem 3.55, for arbitrary values of α, that is, generalize Relations (5.67).

P 5.6. Prove that there exists no sequence transformation of the form

$$y_\nu = \frac{x_{\nu+1} - \gamma \cdot x_\nu}{1-\gamma} ,$$

which inverts the monotonicity direction of the sequence $\{x_\nu\}$, defined through $x_\nu = \frac{1}{\nu}$ for all ν.

6. Generalizations and Final Remarks

A common property of almost all extrapolation procedures studied in this book is that of being *triangular*, i.e., each element of stage k, say $y_i^{(k)}$, can be computed from two others of stage $(k-1)$ through a process of the form

$$y_i^{(k)} = \lambda\, y_{i+1}^{(k-1)} + \mu\, y_i^{(k-1)}; \tag{6.1}$$

here, λ and μ are given coefficients, which may also depend on k and i. If we display these numbers $\{y_i^{(k)}\}$ graphically, we obtain a Romberg tableau or, in other words, a triangular table, which explains why schemes like (6.1) are called triangular.

In this final section we want to embed the extrapolation procedures in the theory of triangular recursion schemes, as developed in C. Brezinski & G. Walz [1].

6.1 Triangular Recursion Schemes

Let us begin with the definition of a triangular recursion scheme; in order to indicate that we leave now the special case of extrapolation procedures and turn to a more general setting, we use a slightly changed notation.

Definition 6.1: Let there be given a sequence $\{T_\nu\}_{\nu\in\mathbb{Z}}$ of elements of a vector space, and define recursively sequences $\{T_\nu^k\}_{\nu\in\mathbb{Z},k\in\mathbb{N}_0}$ as follows:

$$T_\nu^0 := T_\nu \quad \text{for } \nu \in \mathbb{Z} \tag{6.2a}$$

$$T_\nu^k := \lambda_\nu^k T_\nu^{k-1} + \mu_\nu^k T_{\nu+1}^{k-1} \quad \text{for } k \in \mathbb{N} \text{ and } \nu \in \mathbb{Z}\,, \tag{6.2b}$$

where the coefficients λ_ν^k and μ_ν^k are real or complex numbers (which can eventually depend, even non-linearly, on some terms of the sequence $\{T_\nu\}$). Then (6.2) is called *triangular recursion scheme.* Sometimes the indices ν or k will be taken only from subsets of $\mathbb{Z}$ or $\mathbb{N}$, respectively.

It is clear that the extrapolation procedures investigated in Sections 4 and 5 fall into the class of such schemes, but the theory is by far not restricted to extrapolation but covers also quite other topics, such as B-spline functions (polynomial and non-polynomial), Bernstein polynomials, de Casteljau's algorithm, divided differences or formal orthogonal polynomials, to quote only a few; for more information, see C.Brezinski & M.Redivo Zaglia [1], C.Brezinski & G.Walz [1], C.Carstensen, G. Mühlbach & G.Schmidt [1], or G.Walz [5].

We first note some easy observation on schemes like (6.2):

Lemma 6.2: *Let, for some* ν *and* k, T_ν^k *denote the result of a triangular recursion scheme of the form (6.2); then there exist numbers* $\alpha_{i,\nu}^k$ *such that* T_ν^k *can be written as*

$$T_\nu^k = \sum_{i=\nu}^{\nu+k} \alpha_{i,\nu}^k \cdot T_i\,. \tag{6.3}$$

It is

$$\alpha_{\nu,\nu}^k = \lambda_\nu^k \alpha_{\nu,\nu}^{k-1}\,, \qquad \alpha_{\nu+k,\nu}^k = \mu_\nu^k \alpha_{\nu+k,\nu+1}^{k-1}\,, \qquad \textit{and}$$

$$\alpha_{i,\nu}^k = \lambda_\nu^k \alpha_{i,\nu}^{k-1} + \mu_\nu^k \alpha_{i,\nu+1}^{k-1} \quad \textit{for } i = \nu+1,\ldots,\nu+k-1\,,$$

with $\alpha^0_{\nu,\nu} = 1$ *. If, for all values of* k *and* ν *, the sum* $\lambda^k_\nu + \mu^k_\nu$ *does not depend on* ν *, say* $\lambda^k_\nu + \mu^k_\nu = \gamma_k$ *, then the coefficients* $\alpha^k_{i,\nu}$ *satisfy the relation*

$$\alpha_k = \sum_{i=\nu}^{\nu+k} \alpha^k_{i,\nu} = \begin{cases} \prod_{l=1}^{k} \gamma_l, & k \geq 1, \\ 1, & k = 0. \end{cases} \tag{6.4}$$

Conversely, if for all values of k *and* ν *the sum* $\sum_{i=\nu}^{\nu+k} \alpha^k_{i,\nu}$ *does not depend on* ν *, say* $\sum_{i=\nu}^{\nu+k} \alpha^k_{i,\nu} = \alpha_k$ *, then the value* $\lambda^k_\nu + \mu^k_\nu =: \gamma_k$ *is also independent of* ν *and we have*

$$\alpha_k = \begin{cases} \prod_{l=1}^{k} \gamma_l, & k \geq 1, \\ 1, & k = 0. \end{cases} \tag{6.5}$$

Proof. Representation (6.3) as well as (6.4) follow by an easy induction with respect to k. To prove (6.5), we use the relation

$$\sum_{i=\nu}^{\nu+k} \alpha^k_{i,\nu} = \lambda^k_\nu \cdot \sum_{i=\nu}^{\nu+k-1} \alpha^{k-1}_{i,\nu} + \mu^k_\nu \cdot \sum_{i=\nu+1}^{\nu+k} \alpha^{k-1}_{i,\nu+1} ,$$

which follows from (6.2b). Using the assumption on the sum $\sum_{i=\nu}^{\nu+k} \alpha^k_{i,\nu}$, this implies

$$\alpha_k = (\lambda^k_\nu + \mu^k_\nu)\, \alpha_{k-1} , \tag{6.6}$$

which shows that $\lambda^k_\nu + \mu^k_\nu$ is independent of ν . Relation (6.5) now follows directly from (6.6). □

An important tool for the investigation of schemes like (6.2) resp. (6.3) is the so-called reference functional, to be defined as follows:

Definition 6.3: Let Σ denote the space of all functions σ, which are defined on some fixed subset $Z = \{z_i\}_{i\in\mathbb{Z}}$ of $\mathbb{R}$. Then the linear functional on Σ,

$$\mathcal{T}_\nu^k(\sigma) \;:=\; \sum_{i=\nu}^{\nu+k} \alpha_{i,\nu}^k \cdot \sigma(z_i) \tag{6.7}$$

(with $\alpha_{i,\nu}^k$ from (6.3)) is well-defined; we call it the *reference functional* of the recursion (6.2) or the transformation (6.3), respectively.

Of course, putting $T_i := \sigma(z_i)$, we obtain

$$\mathcal{T}_\nu^k(\sigma) \;\equiv\; T_\nu^k \;.$$

The next lemma will be the basis for the proofs of Theorems 6.5 and 6.6, the main results of this section; again, we use the notation

$$\det\begin{pmatrix} f_0 & f_1 & \cdots & f_k \\ y_0 & \cdots & \cdots & y_k \end{pmatrix} \;:=\; \det\begin{pmatrix} f_0(y_0) & \cdots & \cdots & f_0(y_k) \\ \vdots & \vdots & \vdots & \vdots \\ \vdots & \vdots & \vdots & \vdots \\ f_k(y_0) & \cdots & \cdots & f_k(y_k) \end{pmatrix}$$

for functions $f_0,\ldots,f_k$ and numbers $y_0,\ldots,y_k$, see Subsection 3.5.2.

Lemma 6.4: *Let, for some ν and k, $\mathcal{T}_\nu^k$ be a reference functional of the form (6.7); furthermore assume that there exists a $(k+1)-$ dimensional linear subspace*

$$\Sigma^k \;:=\; \operatorname{span}\{\sigma_0,\ldots,\sigma_k\}$$

of Σ, such that

$$\det\begin{pmatrix} \sigma_0 & \sigma_1 & \cdots & \sigma_k \\ z_\nu & \cdots & \cdots & z_{\nu+k} \end{pmatrix} \;\neq\; 0 \tag{6.8}$$

and

$$\mathcal{T}_\nu^k(\sigma_j) \;=\; \begin{cases} \omega_\nu^k\,, & \text{if } j = 0\,, \\ 0\,, & \text{if } j = 1, 2, \ldots, k\,, \end{cases} \tag{6.9}$$

with a certain non-zero number ω_ν^k . *Then every linear functional* L *of the form*

$$L(\sigma) \;:=\; \sum_{i=\nu}^{\nu+k} \beta_{i,\nu}^k \cdot \sigma(z_i)\,, \tag{6.10}$$

which satisfies (6.9), coincides with $\mathcal{T}_\nu^k$. *The space* Σ^k *will be called* <u>*characteristic space*</u> *of* $\mathcal{T}_\nu^k$.

Proof. The difference functional $\Delta := \mathcal{T}_\nu^k - L$ obviously satisfies

$$\Delta(\sigma_j) \;=\; 0 \qquad \text{for } j = 0, \ldots, k\,.$$

This implies $\Delta \equiv 0$, due to the assumed regularity of the above system of equations, and thus $\mathcal{T}_\nu^k \equiv L$. □

Assumption (6.8) is satisfied in particular, if Σ^k is a Haar subspace of Σ . This is indeed the case in many applications.

Theorem 6.5 (C. Brezinski & G. Walz [1]): *Let there be given a transformation* T_ν^k *with reference functional* $\mathcal{T}_\nu^k(\sigma)$. *Then, under the assumptions of Lemma 6.4, the representation*

$$\mathcal{T}_\nu^k(\sigma) \;\equiv\; \frac{\det\begin{pmatrix} \sigma & \sigma_1 & \cdots & \sigma_k \\ z_\nu & \cdots & \cdots & z_{\nu+k} \end{pmatrix}}{\det\begin{pmatrix} \sigma_0 & \sigma_1 & \cdots & \sigma_k \\ z_\nu & \cdots & \cdots & z_{\nu+k} \end{pmatrix}} \cdot \omega_\nu^k, \tag{6.11}$$

is valid.

Proof. The quotient in (6.11) is a linear functional of the form (6.10) and satisfies (6.9). □

Example. In Subection 3.5.1 we had already investigated classical divided differences; let us re-state their recursion rule as follows: With a function f, defined on a set of mutually prime real numbers $\{z_\nu\}$, we define the divided differences D_ν^k of f by

$$D_\nu^0 := f(z_\nu) \qquad \text{for } \nu \in \mathbb{Z} \tag{6.12a}$$

$$D_\nu^k := \frac{D_\nu^{k-1} - D_{\nu+1}^{k-1}}{z_\nu - z_{\nu+k}} \qquad \text{for } k \in \mathbb{N} \text{ and } \nu \in \mathbb{Z}. \tag{6.12b}$$

This is a recursion scheme of the form (6.2); its reference functional $\mathcal{D}_\nu^k$ satisfies

$$\mathcal{D}_\nu^k(p) = \begin{cases} 1, & \text{if } p(z) = z^k , \\ 0 , & \text{if } p \in \Pi_{k-1} . \end{cases}$$

So, in the setting of Lemma 6.4, we have $\Sigma^k = \Pi_k$, the space of polynomials of degree at most k. Now we can apply Theorem 6.5 and get that

$$\mathcal{D}_\nu^k(f) \equiv \frac{\det\begin{pmatrix} f & p_1 & \cdots & p_k \\ z_\nu & \cdots & \cdots & z_{\nu+k} \end{pmatrix}}{\det\begin{pmatrix} p_0 & p_1 & \cdots & p_k \\ z_\nu & \cdots & \cdots & z_{\nu+k} \end{pmatrix}}$$

with $p_j(z) := z^{j-1}$ for $j = 1, \ldots, k$, and $p_0(z) = z^k$; of course, this special case was known before, see Section 3.5 or N.E.NÖRLUND [1].

While it might have been expected that a linear functional like (6.7) can be represented as a quotient of determinants, it is maybe surprising that this can also be done in terms of a complex contour integral:

Theorem 6.6 (C.BREZINSKI & G.WALZ [1]): *Adopt the notations of Lemma 6.4, introduce the abbreviation*

$$A_\nu^k(z) = (z - z_\nu) \cdots (z - z_{\nu+k}) ,$$

and let C *be a simply closed curve in the complex plane, such that the points* $z_\nu, \ldots, z_{\nu+k}$ *lie inside of* C. *Furthermore, assume that the functions* $\sigma, \sigma_0, \ldots, \sigma_k$ *have analytic continuations into the complex plane, and that for* $j = 1, \ldots, k$

$$\frac{|\sigma_j(z)|}{|A_\nu^k(z)|} = o(|z|^{-1}) \quad \text{for } |z| \to \infty . \tag{6.13}$$

Then, if

$$\Omega_\nu^k := \frac{1}{2\pi i} \int_C \frac{\sigma_0(z)}{A_\nu^k(z)} \, dz \neq 0 ,$$

the reference functional $\mathcal{T}_\nu^k$ *has the representation*

$$\mathcal{T}_\nu^k(\sigma) \equiv \frac{1}{2\pi i} \frac{\omega_\nu^k}{\Omega_\nu^k} \int_C \frac{\sigma(z)}{A_\nu^k(z)} \, dz . \tag{6.14}$$

Proof. We denote the integral on the right hand side of (6.14) by I_ν^k. As an immediate consequence of the theorem of residues, I_ν^k is a linear functional of the form

$$I_\nu^k(\sigma) = \sum_{i=\nu}^{\nu+k} \beta_{i,\nu}^k \cdot \sigma(z_i) .$$

We only have to show that Conditions (6.9) are satisfied. For $j = 0$ this is clear. Now let j be a fixed number from the set $\{1, \ldots, k\}$. Because all the poles of the integrand lie inside of C, we can replace this curve by a circle C_R around the origin with radius R, R large enough. This implies the inequality

$$|I_\nu^k(\sigma_j)| \leq \left| \frac{\omega_\nu^k}{\Omega_\nu^k} \right| \cdot R \cdot \left| \frac{\sigma_j(z)}{A_\nu^k(z)} \right| . \tag{6.15}$$

For $R \to \infty$ it follows from (6.13) that the right hand side of (6.15) converges to zero; but this is only possible if

$$I_\nu^k(\sigma_j) \equiv 0 ,$$

which completes the proof of Theorem 6.6. □

Example. We continue the divided difference example from above. For $j = 1, \ldots, k$ it is

$$\frac{|p_j(z)|}{|A_\nu^k(z)|} = O(|z|^{j-k-2}) = o(|z|^{-1}) \quad \text{for } |z| \to \infty ,$$

since $A_\nu^k(z) \in \Pi_{k+1}$. Hence

$$\frac{1}{2\pi i} \int_C \frac{p(z)}{A_\nu^k(z)}\, dz = 0 \tag{6.16}$$

for all $p \in \Pi_{k-1}$. Now let $j = 0$; the function $p_0(z) = z^k$ can be written as

$$z^k = (z - z_{\nu+1})(z - z_{\nu+2}) \cdots (z - z_{\nu+k}) - p(z)$$

with some $p \in \Pi_{k-1}$. From (6.16) we see that

$$\frac{1}{2\pi i} \int_C \frac{z^k}{A_\nu^k(z)}\, dz = \frac{1}{2\pi i} \int_C \frac{1}{z - z_\nu}\, dz \equiv 1 ,$$

and so Theorem 6.6 yields that the reference functional of the divided difference scheme (6.12) can be written as

$$\mathcal{D}_\nu^k(f) = \frac{1}{2\pi i} \int_C \frac{f(z)}{A_\nu^k(z)}\, dz .$$

6.2 A Quite General Extrapolation Algorithm

The first and at the same time main result of this section is given in the following theorem. It says that, under a certain weak assumption on the characteristic space, *every* transformation of the form (6.3) can be computed by a triangular recursion scheme like (6.2), and also gives the expressions for the coefficients λ_ν^k and μ_ν^k. We note already here that this result is so general in scope that it even provides a generalization of the well-known E-algorithm, see below.

Theorem 6.7 (C.BREZINSKI & G.WALZ [1]): *Let, for some* ν *and* k *,*

$$\mathcal{T}_\nu^k(\sigma) = \sum_{i=\nu}^{\nu+k} \alpha_{i,\nu}^k \cdot \sigma(z_i)$$

denote the reference functional of a transformation and let $\Sigma^k :=$ $\operatorname{span}\{\sigma_0, \ldots, \sigma_k\}$ *be its characteristic space. Then, provided that*

$$\det \begin{pmatrix} \sigma_1 & \sigma_2 & \cdots & \sigma_{k-1} \\ z_{\nu+1} & \cdots & \cdots & z_{\nu+k-1} \end{pmatrix} \neq 0 \tag{6.17}$$

(if $k \geq 2$ *), the following recursion scheme holds:*

$$\mathcal{T}_\nu^k(\sigma) = \frac{\mathcal{T}_{\nu+1}^{k-1}(\sigma_k)\mathcal{T}_\nu^{k-1}(\sigma) - \mathcal{T}_\nu^{k-1}(\sigma_k)\mathcal{T}_{\nu+1}^{k-1}(\sigma)}{\omega_\nu^{k-1}\mathcal{T}_{\nu+1}^{k-1}(\sigma_k) - \omega_{\nu+1}^{k-1}\mathcal{T}_\nu^{k-1}(\sigma_k)} \cdot \omega_\nu^k \tag{6.18}$$

with $\mathcal{T}_\nu^0(\sigma) = \sigma(z_\nu)$ *for all* ν *.*

Proof. Let us write

$$\mathcal{T}_\nu^k(\sigma) = \lambda_\nu^k \mathcal{T}_\nu^{k-1}(\sigma) + \mu_\nu^k \mathcal{T}_{\nu+1}^{k-1}(\sigma) . \tag{6.19}$$

For $\sigma = \sigma_0$ this gives us

$$\omega_\nu^k = \lambda_\nu^k \omega_\nu^{k-1} + \mu_\nu^k \omega_{\nu+1}^{k-1} ,$$

while for $\sigma = \sigma_k$ we obtain

$$0 = \lambda_\nu^k \mathcal{T}_\nu^{k-1}(\sigma_k) + \mu_\nu^k \mathcal{T}_{\nu+1}^{k-1}(\sigma_k) .$$

Solving this systems of two equations with two unknowns gives us

$$\lambda_\nu^k = \frac{\omega_\nu^k \cdot \mathcal{T}_{\nu+1}^{k-1}(\sigma_k)}{\omega_\nu^{k-1}\mathcal{T}_{\nu+1}^{k-1}(\sigma_k) - \omega_{\nu+1}^{k-1}\mathcal{T}_\nu^{k-1}(\sigma_k)} \tag{6.20}$$

and

$$\mu_\nu^k = - \frac{\omega_\nu^k \cdot \mathcal{T}_\nu^{k-1}(\sigma_k)}{\omega_\nu^{k-1}\mathcal{T}_{\nu+1}^{k-1}(\sigma_k) - \omega_{\nu+1}^{k-1}\mathcal{T}_\nu^{k-1}(\sigma_k)} . \tag{6.21}$$

Obviously, for $\sigma = \sigma_1, \ldots, \sigma_{k-1}$, (6.18) also holds.

Thus the theorem is proved, if we can show that the denominator in (6.20) and (6.21) is different from zero. To do this, we use (6.11) and write

$$\omega_\nu^{k-1} T_{\nu+1}^{k-1}(\sigma_k) - \omega_{\nu+1}^{k-1} T_\nu^{k-1}(\sigma_k)$$

$$= \omega_\nu^{k-1}\omega_{\nu+1}^{k-1} \cdot \left(\frac{\det\begin{pmatrix} \sigma_k & \sigma_1 & \cdots & \sigma_{k-1} \\ z_{\nu+1} & \cdots & \cdots & z_{\nu+k} \end{pmatrix}}{\det\begin{pmatrix} \sigma_0 & \sigma_1 & \cdots & \sigma_{k-1} \\ z_{\nu+1} & \cdots & \cdots & z_{\nu+k} \end{pmatrix}} - \right.$$

$$\left. - \frac{\det\begin{pmatrix} \sigma_k & \sigma_1 & \cdots & \sigma_{k-1} \\ z_{\nu} & \cdots & \cdots & z_{\nu+k-1} \end{pmatrix}}{\det\begin{pmatrix} \sigma_0 & \sigma_1 & \cdots & \sigma_{k-1} \\ z_{\nu} & \cdots & \cdots & z_{\nu+k-1} \end{pmatrix}} \right) ,$$

which is, by Sylvester's determinantal identity, for $k \geq 2$ equal to

$$(-1)^k \omega_\nu^{k-1}\omega_{\nu+1}^{k-1} \cdot$$

$$\cdot \frac{\det\begin{pmatrix} \sigma_0 & \sigma_1 & \cdots & \sigma_k \\ z_{\nu} & \cdots & \cdots & z_{\nu+k} \end{pmatrix} \det\begin{pmatrix} \sigma_1 & \sigma_2 & \cdots & \sigma_{k-1} \\ z_{\nu+1} & \cdots & \cdots & z_{\nu+k-1} \end{pmatrix}}{\det\begin{pmatrix} \sigma_0 & \sigma_1 & \cdots & \sigma_{k-1} \\ z_{\nu} & \cdots & \cdots & z_{\nu+k-1} \end{pmatrix} \det\begin{pmatrix} \sigma_0 & \sigma_1 & \cdots & \sigma_{k-1} \\ z_{\nu+1} & \cdots & \cdots & z_{\nu+k} \end{pmatrix}} .$$

Since we always assume that

$$\det\begin{pmatrix} \sigma_0 & \sigma_1 & \cdots & \sigma_k \\ z_{\nu} & \cdots & \cdots & z_{\nu+k} \end{pmatrix} \neq 0 ,$$

the additional assumption (6.17) gives the desired result.
In the case $k = 1$, the denominator in (6.20) and (6.21) reads

$$\omega_\nu^0 T_{\nu+1}^0(\sigma_1) - \omega_{\nu+1}^0 T_\nu^0(\sigma_1) ,$$

which equals

$$\det\begin{pmatrix} \sigma_0 & \sigma_1 \\ z_{\nu} & z_{\nu+1} \end{pmatrix} ,$$

a non-zero number. Thus the theorem is proved. □

Probably the most general extrapolation algorithm known so far is the E-algorithm, which covers almost all existing extrapolation methods as special cases. The E-algorithm was discovered independently (and using different approaches) by T.HÅVIE [4] and C.BREZINSKI [2]; for special cases, it was known before, see e.g. C.SCHNEIDER [1]. Meanwhile, the E-algorithm is well-described in many publications, such as J.WIMP [1] or, in particular, C.BREZINSKI & M.REDIVO ZAGLIA [1], which also includes an extensive list of references; moreover, a new approach can be found in C.BREZINSKI & M.REDIVO ZAGLIA [2].

The connection to our investigations is given by the maybe surprising fact that the algorithm (6.18) is once more a generalization of the E-algorithm, which itself is recovered by setting

$$\begin{aligned} E_k^{(\nu)} &= \mathcal{T}_\nu^k(\sigma)/\mathcal{T}_\nu^k(\sigma_0) \quad \text{and} \\ g_{k,i}^{(\nu)} &= \mathcal{T}_\nu^k(\sigma_i)/\mathcal{T}_\nu^k(\sigma_0) \,. \end{aligned}$$

The E-algorithm is a very genaral tool and has without any doubt many important applications in several branches of numerical analysis; but we allow us at this place once more to remind the reader of the fundamental philosophy of this book, which is: Keep the extrapolation process you want to use as simple as possible, such that the effect of convergence acceleration is not disturbed by a maybe too complicated extrapolation procedure. This leads us almost automatically to linear extrapolation, which, in addition, can also be applied to vector or matrix sequences without difficulties. However, readers are encouraged to study the E-algorithm in one of the above-mentioned publications.

As an *example* for the situation of Theorem 6.7, let us have a look at the *quasi-linear extrapolation methods*:

Let $\{T_\nu\}$ be a sequence of numbers. We say that the transformation $\{T_\nu\} \to \{T_\nu^k\}_{\nu\in\mathbb{N}}$ is quasi-linear, if it has the form

$$T_\nu^k \;:=\; F_k(T_\nu,\ldots,T_{\nu+k}) \,,$$

where the function F_k satisfies

$$F_k(au_0+b,\ldots,au_k+b) \;=\; a\,F_k(u_0,\ldots,u_k)+b$$

for fixed numbers a and b. In that case, as proved in C.BREZINSKI [5], we have

$$T_\nu^k \;=\; \sum_{i=\nu}^{\nu+k} \alpha_{i,\nu}^k \cdot T_i \,,$$

where, for $i = \nu, \ldots, \nu + k$,

$$\alpha_{i,\nu}^{k} = F'_{k,i}(T_\nu, \ldots, T_{\nu+k})$$

with

$$F'_{k,i}(u_0, \ldots, u_k) = \frac{\delta F_k(u_0, \ldots, u_k)}{\delta u_{i-\nu}} .$$

Let us examine a familiar example for quasi-linear extrapolation methods in detail: Aitken's Δ^2-process (cf. Section 4.1), which can be written as

$$\begin{aligned} T_\nu^1 &= \frac{S_\nu \Delta S_{\nu+1} - S_{\nu+1}\Delta S_\nu}{\Delta S_{\nu+1} - \Delta S_\nu} \\ &= \lambda_\nu^1 S_\nu + \mu_\nu^1 S_{\nu+1} \end{aligned}$$

with

$$\lambda_\nu^1 = \frac{\Delta S_{\nu+1}}{\Delta^2 S_\nu} \quad \text{and} \quad \mu_\nu^1 = -\frac{\Delta S_\nu}{\Delta^2 S_\nu} .$$

Thus, if $g = \{g(\nu)\}$, we have

$$T_\nu^1(g) = \frac{g(\nu)\Delta S_{\nu+1} - g(\nu+1)\Delta S_\nu}{\Delta S_{\nu+1} - \Delta S_\nu} .$$

So, the two sequences defined by $g_0(\nu) = 1$ and $g_1(\nu) = \Delta S_\nu$ satisfy our assumptions, since for all ν

$$T_\nu^1(g_0) = 1 \quad \text{and} \quad T_\nu^1(g_1) = 0 ,$$

and we obtain the representation

$$T_\nu^1 = \frac{\det\begin{pmatrix} S_\nu & S_{\nu+1} \\ \Delta S_\nu & \Delta S_{\nu+1} \end{pmatrix}}{\det\begin{pmatrix} 1 & 1 \\ \Delta S_\nu & \Delta S_{\nu+1} \end{pmatrix}} ,$$

which is a well-known formula for Aitken's Δ^2-process.

We now conclude our investigations by presenting the following "multistep formula":

Theorem 6.8 (C.BREZINSKI & G.WALZ [1]): *Let, for some $m \geq 1$, the determinant*

$$\det\begin{pmatrix} \mathcal{T}_\nu^k & \mathcal{T}_{\nu+1}^k & \cdots & \mathcal{T}_{\nu+m}^k \\ \sigma_0 & \sigma_{k+1} & \cdots & \sigma_{k+m} \end{pmatrix}$$

$$:= \det\begin{pmatrix} \mathcal{T}_\nu^k(\sigma_0) & \mathcal{T}_\nu^k(\sigma_{k+1}) & \cdots & \mathcal{T}_\nu^k(\sigma_{k+m}) \\ \mathcal{T}_{\nu+1}^k(\sigma_0) & \mathcal{T}_{\nu+1}^k(\sigma_{k+1}) & \cdots & \mathcal{T}_{\nu+1}^k(\sigma_{k+m}) \\ \vdots & \vdots & \vdots & \vdots \\ \mathcal{T}_{\nu+m}^k(\sigma_0) & \mathcal{T}_{\nu+m}^k(\sigma_{k+1}) & \cdots & \mathcal{T}_{\nu+m}^k(\sigma_{k+m}) \end{pmatrix}$$

be different from zero. Then the value $\mathcal{T}_\nu^{k+m}(\sigma)$ can be expressed as linear combination of the values $\mathcal{T}_\nu^k(\sigma), \ldots, \mathcal{T}_{\nu+m}^k(\sigma)$. We have the relation

$$\mathcal{T}_\nu^{k+m}(\sigma) = \frac{\det\begin{pmatrix} \mathcal{T}_\nu^k & \mathcal{T}_{\nu+1}^k & \cdots & \mathcal{T}_{\nu+m}^k \\ \sigma & \sigma_{k+1} & \cdots & \sigma_{k+m} \end{pmatrix}}{\det\begin{pmatrix} \mathcal{T}_\nu^k & \mathcal{T}_{\nu+1}^k & \cdots & \mathcal{T}_{\nu+m}^k \\ \sigma_0 & \sigma_{k+1} & \cdots & \sigma_{k+m} \end{pmatrix}} \cdot \omega_\nu^{k+m}, \tag{6.22}$$

and a similar formula by exchanging k and m.

Proof. Our aim is to deduce an expression of the form

$$\mathcal{T}_\nu^{k+m}(\sigma) = a_0 \mathcal{T}_\nu^k(\sigma) + \cdots + a_m \mathcal{T}_{\nu+m}^k(\sigma)$$

for arbitrary $m \geq 1$ (where the coefficients $a_0, \ldots, a_m$ do of course also depend on ν and k) Taking here successively $\sigma = \sigma_0, \sigma_{k+1}, \ldots, \sigma_{k+m}$ we obtain the coefficients $a_0, \ldots, a_m$ as solutions of the linear system

$$T \cdot \begin{pmatrix} a_0 \\ \vdots \\ \vdots \\ a_m \end{pmatrix} = \begin{pmatrix} \omega_\nu^{k+m} \\ 0 \\ \vdots \\ 0 \end{pmatrix}$$

with

$$T := \begin{pmatrix} \mathcal{T}_\nu^k(\sigma_0) & \mathcal{T}_{\nu+1}^k(\sigma_0) & \cdots & \mathcal{T}_{\nu+m}^k(\sigma_0) \\ \mathcal{T}_\nu^k(\sigma_{k+1}) & \mathcal{T}_{\nu+1}^k(\sigma_{k+1}) & \cdots & \mathcal{T}_{\nu+m}^k(\sigma_{k+1}) \\ \vdots & \vdots & \vdots & \vdots \\ \mathcal{T}_\nu^k(\sigma_{k+m}) & \mathcal{T}_{\nu+1}^k(\sigma_{k+m}) & \cdots & \mathcal{T}_{\nu+m}^k(\sigma_{k+m}) \end{pmatrix},$$

and thus, by Cramer's rule, the assertion of the theorem. □

For $k = 0$ we recover the determinantal formula (6.11) while, for $m = 1$, we obtain (6.18). For an arbitrary value of m this multistep formula can be used to compute directly $\mathcal{T}_\nu^{k+m}(\sigma)$ from the $\mathcal{T}_\nu^k(\sigma)$'s. Thus if a breakdown, due to a division by zero, occurs in (6.18), it is possible to jump over the singularity. The same is true if the denominator in (6.18) is small in absolute value. In this case, called near-breakdown, a severe propagation of rounding errors can occur and thus using (6.22) can improve the numerical stability of the algorithm by jumping over the near-singularity.

6.3 Problems

P 6.1 (G.Walz [5]). Apply Theorem 6.6 to the process of logarithmic extrapolation, i.e., write the number $y_i^{(k)}$, defined through (4.31), as a contour integral.

P 6.2. Prove that, in the situation of Theorem 6.6, the coefficients $\alpha_{i,\nu}^k$ in the representation

$$\mathcal{T}_\nu^k(\sigma) = \sum_{i=\nu}^{\nu+k} \alpha_{i,\nu}^k \cdot \sigma(z_i)$$

of the reference functional $\mathcal{T}_\nu^k$ in (6.14) are explicitly given by

$$\alpha_{i,\nu}^k = \frac{\omega_\nu^k}{\Omega_\nu^k} \cdot \left(\frac{d}{dz} A_\nu^k(z)_{z=z_i}\right)^{-1}.$$

Historical Notes

It was already indicated several times in this book that the fundamental ideas of extrapolation can be traced back to publications, which appeared long time before the papers by Richardson/Gaunt and Romberg. On these final pages we want to make some notes on this development, beginning with Archimedes end ending with Romberg, whose nowadays famous paper W.ROMBERG [1] gave, without any doubt, the initialization of modern extrapolation theory (where, of course, until Romberg's work "extrapolation " was a synonym for "logarithmic extrapolation in the standard case with $b = 2$ and $n_0 = 1$"). Also, readers who are interested in some additional information on the history of extrapolation methods, are kindly referred to the papers by C.BREZINSKI [4,6], J.DUTKA [1], H.ENGELS [2], D.C.JOYCE [1], or G.WALZ [11].

When looking at the development of extrapolation theory in the last centuries, it is remarkable that most of the authors did not take notice of each others work, which led to a significant amount of rediscovering. In particular, this is true for the two groups of people working on the computation of π on one side, and on the numerical integration of differential equations and functions on the other, which developed over the centuries very similar numerical methods for the solution of their respective problem without knowing each other's work. To give only one example, Richardson's famous h^2- extrapolation for the numerical solution of ODEs, developed in 1927, is nothing else but Huygens' method for the numerical computation of π, which was found in 1654 (see below).

In some sense, the story begins with the work of ARCHIMEDES [1] in 250 B.C. Although Archimedes did not work on the field of extrapolation methods itself, his name must be quoted in this survey, since he founded a method for the approximative computation of π, which later on gave rise for the very important work of Huygens, Richardson/Gaunt, Kommerell and others; we had described Archimedes' approach in Subsections 3.1.1 and 5.2.3.

For many centuries this was the most widespread method for the computation of π, and the only progress that was made consisted in increasing the number of vertices of the respective polygons. For example, Ludolf van Ceulen (= from Cologne) in the year 1610 obtained 35 digits of π by calculating the area A_n of the regular polygon with $n = 2^{62}$ vertices!

The first methodical advance is due to CH.HUYGENS [1] in 1654; using geometrical arguments, he showed that the sequence $\{T_n\}$, defined by

$$T_n := (4\,A_{2n} - A_n)/3$$

converges faster to the limit π than the sequence $\{A_n\}$ itself does. So, Huygens found out the first step (but only the first) of the extrapolation procedure (4.31) as a convergence accelerating method in this special case, see again Subsection 5.2.3.

The next milestone in the history of extrapolation methods is, without any doubt, the booklet of J.F.SAIGEY [1] from 1859, which unfortunately was overlooked for more than a century and was rediscovered in 1984 by J.DUTKA [1]. Saigey developed, with purely analytical methods, in particular without refering to Archimedes or Huygens, the existence of an asymptotic expansion for the sequence $\{A_n\}$, i.e. he proved (with a slight change of notations)

$$A_n = \pi + \frac{c_1}{n^2} + \frac{c_2}{n^4} + \frac{c_3}{n^6} + \cdots ,$$

where the c_ν are fixed coefficients, and then derived from this relation his "higher approximations" to π, which turn out to be nothing else than the results of the logarithmic extrapolation process: Considering $\{A_n\}$ as the sequence of *first approximations*, he defines the *second approximations*,

$$\tilde{A}_n := A_{2n} + \frac{1}{3} \cdot (A_{2n} - A_n) ,$$

the *third approximations*,

$$B_n := \tilde{A}_{2n} + \frac{1}{15} \cdot (\tilde{A}_{2n} - \tilde{A}_n) ,$$

the *fourth approximations*,

$$C_n := B_{2n} + \frac{1}{63} \cdot (B_{2n} - B_n) ,$$

and so on. So, Saigey was the very first who developed a special case of the extrapolation process (4.31) in *iterative form*; note that this was almost precisely 100 years before Rombergs paper appeared.

In a historical note, Saigey remarked that he first had made a wrong conjecture concerning the denominators in his higher approximations: after having noted that the first ones equal 3 and 15, he had conjectured that the rule for constructing these denominators is

$$1 \cdot 3 = 3, \quad 1 \cdot 3 \cdot 5 = 15, \quad 1 \cdot 3 \cdot 5 \cdot 7 = 105, \ \ldots$$

and so on, which is *wrong*. In a private communication, the engineer and geometer Guerin pointed out to him the correct rule, which is

$$4^1 - 1 = 3, \quad 4^2 - 1 = 15, \quad 4^3 - 1 = 63, \ \ldots .$$

Now we leave for a moment the problem of computing π and turn to the numerical integration of ordinary differential equations and functions.

The first important work in this context is the famous "Deferred Approach to the Limit" from 1927, a paper which consists of two parts; the first one ("Single Lattice") is due to

L.F.RICHARDSON [3], the second one ("Interpenetrating Lattices") to J.A.GAUNT [1]. Usually only the first part is quoted, and people therefore speak of *Richardson-extrapolation.* The method of Richardson and Gaunt (Subsection 4.3.2) is – also building on the work of Archimedes – a precise transcription of Huygens' method, which they obviously were not aware of, to the problem of integrating ordinary differential equations. They called it "h^2– extrapolation", since they treated special logarithmic asymptotic expansions with $\rho_\mu = 2\mu$ for all μ, and eliminated just the first ($= h^2-$) term of the expansion. It should be noticed that Richardson himself referred to N.BOGOLJUBOV & N.KRYLOV [1], in which the "deferred approach to the limit" can already be found (information from H.ENGELS [2]). Just like Huygens before, Richardson and Gaunt did not try to bring their method into an iterative form, i.e. they did not want to eliminate further terms of the asymptotic expansion. If only they had...!

Another very important publication on extrapolation methods, which unfortunately seems to have been overlooked for all the years, is that of K.KOMMERELL [1]. In spite of the fact that this book is referred to in the nice surveys by H.ENGELS [2] and J.DUTKA [1], I think that still not all the deep ideas of Kommerell are exploited; the most important one in the context of asymptotic expansion theory was investigated in Subsection 5.2.3.

The great importance of his work lies in the fact that it already contains the *recursive extrapolation algorithm* (4.31), applied to Archimedes' sequence $\{A_n\}$ (K.KOMMERELL [1], p. 37)! So, what Kommerell did is – just like Saigey – nothing else but to write down Romberg's method, here in the context of approximating π.

And there is also another aspect which connects Kommerell's work with the one of Romberg, namely the fact that both define iteratively *two* approximating sequences, which *include* the limit value π resp. the exact value of the integral, cf. again Subsection 5.2.3.

Of course, the great breakthrough of extrapolation methods must indeed be dated to the year 1955, when Romberg's fundamental and in the meantime famous paper W.ROMBERG [1] on "Vereinfachte numerische Integration" (simplified numerical integration) appeared. As already pointed out, many authors denote the extrapolation procedure (4.31) also as "generalized Romberg method" or just speak of "Romberg's principle". Although Romberg himself never thought of constructing a method for extrapolation or of giving a general principle for convergence acceleration (he "only" wanted to find new formulae for numerical integration), these names are justified inasmuch, as his paper inspired very many authors to work on this field and produce some nice results, which nowadays influence many parts of numerical analysis; some of the most important contributions to the theory we had faced troughout this book.

Finally, I would like to point out once more that the history of extrapolation methods was over the years characterized by a lot of disregarding and re-discovering, what is quite a pity. Not only that many energy could have been saved, if the authors had taken notice of each others work, also some of the methods bear – from a historical point of view – unsatisfactory names; for example, the Richardson-extrapolation process could have been denoted as *Huygens-extrapolation*, whereas Romberg's method maybe should have been named after *Kommerell* or *Saigey*. However, in the sense of the above-made remarks on the importance of Romberg's work, all these names are maybe nevertheless good as they are.

References

M. ABRAMOWITZ & I. STEGUN.

[1] *Handbook of mathematical functions.* Dover Publications, New York 1965

J. ALBRECHT.

[1] *Intervallschachtelungen beim Romberg-Verfahren.* Z. Ang. Math. Mech. **52** (1972), 433 – 435

A. L. ANDREW.

[1] *Convergence of an iterative method for derivatives of eigensystems.* J. Comp. Phys. **26** (1978), 107 – 112

[2] *Iterative computation of derivatives of eigenvalues and eigenvectors.* J. Inst. Math. Appl. **24** (1979), 209 – 218

H. ANTES.

[1] *Über die vierdimensionale Romberg-Integration mit Schranken.* Computing **9** (1972), 45 – 52

[2] *Über die Romberg-Integration in n Dimensionen.* Computing **9** (1972), 203 – 209

ARCHIMEDES.

[1] *κυκλου μετρησις.* A translation (German) can be found in: F. RUDIO, *Archimedes, Huygens, Lambert, Legendre – Vier Abhandlungen über Kreismessung.* Teubner, Leipzig 1892, 71 – 81

W. AUZINGER.

[1] *On the error structure of the implicit Euler scheme applied to stiff systems of differential equations.* Computing **43** (1989), 115 – 131

[2] *On error structures and extrapolation for stiff systems, with application in the method of lines.* Computing **44** (1990), 331 – 356

W. AUZINGER & R. FRANK.

[1] *Asymptotic expansions of the global discretization error for stiff problems.* SIAM J. Sci. Stat. Comp. **10** (1989), 950 – 963

C. T. BAKER & G. S. HODGSON.

[1] *Asymptotic expansions for integration formulas in one or more dimensions.* SIAM J. Numer. Anal. **8** (1971), 473 – 480

W. Baron.

[1] *Über eine neue Zerlegungsfolge bei der Richardson-Extrapolation.* Arch. Math. **30** (1978), 443-448

F. L. Bauer.

[1] *La méthode de l'intégration numérique de Romberg.* Colloque sur l'Analyse Numérique, Libraire Universitaire, Louvain 1961, 119 – 129

F. L. Bauer, H. Rutishauser & E. Stiefel.

[1] New aspects in numerical quadrature. In: *Proc. Symp. Applied Mathematics, Vol. 15.* American Mathematical Society, Providence, R. I 1963, 199 – 218

Jakob Bernoulli.

[1] *Ars Conjectandi.* Basel 1713

S. N. Bernstein.

[1] *Sur la meilleure approximation de $|x|$ par des polynomes de degrés donnés.* Acta Math. **37** (1914), 1 – 57

P. Bjørstadt, G. Dahlquist & E. Grosse.

[1] *Extrapolation of asymptotic expansions by a modified Aitken δ^2- formula.* BIT **21** (1981), 56 – 65

H. Blum.

[1] *Asymptotic error expansion and defect correction in the finite element method.* Technical Report 640, SFB 123, University of Heidelberg, 1991

H. Blum, Q. Lin & R. Rannacher.

[1] *Asymptotic error expansion and Richardson extrapolation for linear finite elements.* Numer. Math. **49** (1986), 11 - 37

H. Blum & R. Rannacher.

[1] *Extrapolation techniques for reducing the pollution effect of reentrant corners in the finite element method.* Numer. Math. **52** (1988), 539 – 564

[2] *Finite element eigenvalue computation on domains with reentrant corners using Richardson extrapolation.* J. Comp. Maths. **8** (1990), 321 – 332

K. Böhmer.

[1] *Fehlerasymptotik von Diskretisierungsverfahren und ihre numerische Anwendung.* Report 77/2, University of Karlsruhe 1977

[2] *Asymptotic error expansions and discrete Newton methods for elliptic boundary value problems.* In: *Numerical integration of differential equations and large linear systems* (Lecture Notes Math. 968), J. Hinze (ed.), Springer, Berlin/Heidelberg 1982, 292 – 300

K. Böhmer, P. Hemker & H. J. Stetter.

[1] *The defect correction approach.* Computing Suppl. **5** (1984), 1 – 32

K. BÖHMER & J. LOCKER.

[1] *Asymptotic expansions for the discretizaton error of least squares solutions of linear boundary value problems.* Math. Comp. **51** (1988), 75 – 91

N. BOGOLJUBOV & N. KRYLOV.

[1] *On Rayleigh's principle in the theory of differential equations of mathematical physics and upon Euler's method in the calculus of variations* (Ukr.). Notices Acad. Sci. Ukraine (Phys. Math.) **3** (1926), 37 – 57

H. C. BOLTON & H. I. SCOINS.

[1] *Eigenvalues of differential equations by finite-difference methods.* Proc. Cambridge Phil. Soc. **52** (1956), 215 – 229

[2] *Eigenvalue problems treated by finite difference methods II.* Proc. Cambridge Phil. Soc. **53** (1957), 150 – 161

H. BRASS.

[1] *Interpolation und $E_n(f)-$ Abschätzung.* In: *Numerische Methoden der Approximationstheorie* **3** (ISNM 30), L. COLLATZ, H. WERNER, G. MEINARDUS (eds.), Birkhäuser, Basel 1976, 91 – 108

[2] *Quadraturverfahren.* Vandenhoeck & Ruprecht, Göttingen 1977

C. BREZINSKI.

[1] *Accéleration de la convergence en analyse numérique* (Lecture Notes Math. 584). Springer, Berlin 1977

[2] *A general extrapolation algorithm.* Numer. Math. **35** (1980), 175 – 187

[3] *Error control in convergence acceleration processes.* IMA J. Numer. Anal. **3** (1983), 65 – 80

[4] *Convergence acceleration methods: The past decade.* J. Comp. Appl. Maths. **12&13** (1985), 19 – 36

[5] *Quasi-linear extrapolation processes.* In: *Numerical mathematics Singapore 1988* (ISNM 86), R. P. AGARVAL et al. (eds.), Birkhäuser, Basel 1988, 61 – 78

[6] *A survey of iterative extrapolation by the E-algorithm.* Det Kong. Norske Vid. Selskab Skr. **2** (1989), 1 – 26

C. BREZINSKI & M. REDIVO ZAGLIA.

[1] *Extrapolation methods. Theory and practice.* North Holland, Amsterdam 1992

[2] *A general extrapolation prodecure revisited.* Adv. in Comp. Maths. **2** (1994), 461 – 477

C. BREZINSKI & G. WALZ.

[1] *Sequences of transformations and triangular recursion schemes, with*

applications in numerical analysis. J. Comp. Appl. Maths. **34** (1991), 361 – 383

N. G. DE BRUIJN.

[1] *Asymptotic methods in analysis* (3rd ed.). North-Holland, Amsterdam/London 1970

R. BULIRSCH.

[1] *Bemerkungen zur Romberg-Integration.* Numer. Math. **6** (1964), 6 – 16

R. BULIRSCH & J. STOER.

[1] *Fehlerabschätzungen und Extrapolation mit rationalen Funktionen bei Verfahren vom Richardson-Typus.* Numer. Math. **6** (1964), 413 – 427

[2] *Numerical treatment of ordinary differential equations by extrapolation methods.* Numer. Math. **8** (1966), 1 – 13

[3] *Asymptotic upper and lower bounds for results of extrapolation methods.* Numer. Math. **8** (1966), 93 – 104

[4] *Asymptotische Fehlerschranken bei Extrapolationsverfahren.* In: *Funktionalanalysis, Approximationstheorie, Numerische Mathematik* (ISNM 7), L. COLLATZ, G. MEINARDUS, H. UNGER (eds.), Birkhäuser, Basel 1967, 165 – 170

[5] *Numerical quadrature by extrapolation.* Numer. Math. **9** (1967), 271 – 278

J. C. BUTCHER.

[1] *On the attainable order of Runge-Kutta methods.* Math. Comp. **19** (1965), 408 – 417

[2] *On the convergence of numerical solutions to ordinary differential equations.* Math. Comp. **20** (1966), 1-10

[3] *The non-existence of ten stage eight order explicit Runge-Kutta methods.* BIT **25** (1985), 521 – 540

C. CARSTENSEN, G. MÜHLBACH & G. SCHMIDT.

[1] *De Casteljau's algorithm is an extrapolation method.* Computer Aided Geom. Design **12** (1995), 371 – 380

A. S. CAVARETTA, C. A. MICCHELLI & A. SHARMA.

[1] *Multivariate interpolation and the Radon transform.* Math. Z. **174** (1980), 263 – 279

A. CAYLEY.

[1] *An elementary treatise on elliptic functions.* Dover, New York 1961

P. C. CHAKRAVARTI.

[1] *Integrals and Sums.* Athlone Press, London 1970

K. CHANDRASEKHARAN.
[1] *Elliptic functions.* Springer, Heidelberg 1985

C. CHIRUTA.
[1] *Analysis of certain possibility for parallel treatment of the Romberg quadrature algoritm.* An. Stiint. Univ. Al. I. Cuza Iasi, Sect. Ia, **28** (Suppl.) (1982), 101 - 104

E. CHRISTIANSEN & H. G. PETERSEN.
[1] *Estimation of convergence orders in repeated Richardson extrapolation.* BIT **29** (1989), 48 - 59

L. COLLATZ.
[1] *Numerische Behandlung von Differentialgleichungen.* Springer, Berlin 1951

J. B. CONWAY.
[1] *Functions of one complex variable* (2nd ed.). Springer, New York/ Berlin/Heidelberg 1978

G. DAHLQUIST.
[1] *Convergence and stability in the numerical integration of ordinary differential equations.* Math. Scand. **4** (1956), 33 - 53

[2] *Stability and error bounds in the numerical integration of ordinary differential equations.* Thesis, Stockholm 1958. Also in: Trans. Royal Inst. Techn. **130**, Stockholm 1959

[3] *A special stability problem for linear multistep methods.* BIT **3** (1963), 27 - 43

[4] *33 years of numerical instability I.* BIT **25** (1985), 188 - 204

G. DAHLQUIST & B. LINDBERG.
[1] *On some implicit one-step methods for stiff differential equations.* Report TRITA-NA-7302, Royal Institute of Technology, Stockholm 1973

P. J. DAVIS & PH. RABINOWITZ.
[1] *Methods of numerical integration* (2nd ed.). Academic Press, New York 1984

J. P. DELAHAYE.
[1] *Sequence transformations.* Springer, Berlin 1988

P. DEUFLHARD.
[1] *Order and stepsize control in extrapolation methods.* Numer. Math. **41** (1983), 399 - 422

[2] *Recent progress in extrapolation methods for ordinary differential equations.* SIAM Rev. **27** (1985), 505 - 535

P. Deuflhard, E. Hairer & J. Zugck.

[1] *One-step and extrapolation methods for differential-algebraic systems.* Numer. Math. **51** (1987), 501 – 516

E. de Doncker.

[1] *New Euler-Maclaurin expansions and their application to quadrature over the s– dimensional simplex.* Math. Comp. **33** (1979), 1003 – 1018

J. E. Drummond.

[1] *Summing a common type of slowly convergent series of positive terms.* J. Austral. Math. Soc., Ser. B **19** (1976), 416 – 421

[2] *Convergence speeding, convergence and summability.* J. Comp. Appl. Maths. **11** (1984), 145 – 159

J. Dutka.

[1] *Richardson-extrapolation and Romberg-integration.* Historia Mathematica **11** (1984), 3 – 21

H. Engels.

[1] *Zur Anwendung der Richardson-Extrapolation auf die numerische Differentiation.* Computing **8** (1971), 255 – 271

[2] *Zur Geschichte der Richardson-Extrapolation.* Historia Mathematica **6** (1979), 280 – 293

A. Erdélyi.

[1] *Asymptotic expansions.* Dover Publications, New York 1956

L. Euler.

[1] *Methodus generalis summandi progressiones.* Comment. Acad. Sc. Petro. **6** (1732/33), 68 – 97

D. J. Evans & G. M. Megson.

[1] *Romberg integration using systolic arrays.* Parallel Comput. **3** (1986), 289 – 304

D. J. Evans & B. B. Sanugi.

[1] *A comparison of numerical ODE solvers based on arithmetic and geometric means.* Intern. J. Computer Math. **23** (1987), 37 – 62

[2] *A comparison of nonlinear trapezoidal formulae for solving initial value problems.* Intern. J. Computer Math. **41** (1991), 65 – 79

[3] *Numerical solution of initial value problems by nonlinear trapezoidal formulae and rational extrapolation methods II.* Preprint, 1994

D. J. Evans & G. Walz.

[1] *Numerical solution of initial value problems by nonlinear trapezoidal formulae and extrapolation methods.* Intern. J. Computer Math. **53** (1994), 83 – 98

G. Fairweather & A. V. Saylor.

[1] *On the application of extrapolation, deferred correction and defect correction to discrete-time Galerkin methods for parabolic problems.* IMA J. Numer. Anal. **3** (1983), 173 – 192

H. Fiedler.

[1] *On the asymptotic behavior of the best L_∞ -approximation by polynomials.* Constr. Approx. **3** (1987), 377 – 388

S. Filippi.

[1] *Die Berechnung einiger elementarer transzendenter Funktionen mit Hilfe des Richardson-Algorithmus.* Computing **1** (1966), 127 – 132

K.-J. Förster.

[1] *Fehlerschranken bei der Romberg-Quadratur.* Z. Ang. Math. Mech. **62** (1982), 133 – 135

[2] *Zum Vergleich von Quadraturformeln.* Dissertation, University of Braunschweig 1982

L. Fox.

[1] *Romberg integration for a class of singular integrands.* Computer J. **10** (1967), 87 – 93

L. Fox & L. Hayes.

[1] *On the definite integration of singular integrands.* SIAM Rev. **12** (1970), 449 – 457

M. Gasca & A. López-Carmona.

[1] *A general recurrence interpolation formula and its application to multivariate interpolation.* J. Approx. Theory **34** (1982), 361 – 374

M. Gasca & J. I. Maeztu.

[1] *On Lagrange and Hermite interpolation in $\mathbb{R}^k$.* Numer. Math. **39** (1982), 1 – 14

J. A. Gaunt.

[1] *The deferred approach to the limit. Part II: Interpenetrating lattices.* Phil. Trans. Royal Soc. London, Ser. A **226** (1927), 350 – 361

C. W. Gear.

[1] *Numerical initial value problems in ordinary differential equations.* Prentice-Hall, Englewood Cliffs 1971

M. de Gee.

[1] *Asymptotic expansions for the midpoint rule applied to delay differential equations.* SIAM J. Numer. Anal. **23** (1986), 1254 – 1272

K. GEORG.

[1] *Approximation of integrals for boundary element methods.* SIAM J. Sci. Stat. Comp. **12** (1991), 443 – 453

S. GONZALEZ-PINTO, P. GONZALEZ-VERA & J. C. SANTOS.

[1] *On extrapolation of Jagerman and Stetter rules.* Numer. Algorithms **3** (1992), 211 – 222

W. B. GRAGG.

[1] *Repeated extrapolation to the limit in the numerical solution of ordinary differential equations.* Thesis, Univ. California, Los Angeles 1964

[2] *On extrapolation algorithms for ordinary initial value problems.* SIAM J. Num. Anal. **2** (1965), 384 – 403

R. D. GRIGORIEFF.

[1] *Numerik gewöhnlicher Differentialgleichungen, Band 1: Einschrittverfahren.* B. G. Teubner, Stuttgart 1972

[2] *Numerik gewöhnlicher Differentialgleichungen, Band 2: Mehrschrittverfahren.* B. G. Teubner, Stuttgart 1977

K. P. HADELER & G. MEINARDUS.

[1] *On the roots of Cauchy polynomials.* Linear Alg. Appl. **38** (1981), 81 – 102

E. HAIRER & CH. LUBICH.

[1] *Asymptotic expansions of the global error of fixed-stepsize methods.* Numer. Math. **45** (1984), 345 – 360

[2] *Extrapolation at stiff differential equations.* Numer. Math. **52** (1988), 377 – 400

E. HAIRER, CH. LUBICH & M. ROCHE.

[1] *The numerical solution of differential-algebraic systems by Runge-Kutta methods* (Lecture Notes Math. 1409). Springer, Berlin/Heidelberg 1989

E. HAIRER, S. P. NØRSETT & G. WANNER.

[1] *Solving ordinary differential equations I: Nonstiff problems.* Springer, Berlin/Heidelberg 1987

E. HAIRER & G. WANNER.

[1] *Solving ordinary differential equations II: Stiff and differential-algebraic problems.* Springer, Berlin/Heidelberg 1991

H. A. HAKOPIAN.

[1] *Multivariate divided differences and multivariate interpolation of Lagrange and Hermite type.* J. Approx. Theory **34** (1982), 286 – 305

A. HAUTOT.

[1] *Convergence acceleration of continued fractions of Poincaré's Type.* Appl. Num. Maths. **4** (1988), 309 – 322

T. HÅVIE.

[1] *On a modification of Romberg's algorithm.* BIT **6** (1966), 24 – 30

[2] *On the practical application of the modified Romberg algorithm.* BIT **7** (1967), 103 – 113

[3] *Derivation of explicit expressions for error terms in the ordinary and modified Romberg algorithms.* BIT **9** (1969), 18 – 29

[4] *Generalized Neville type extrapolation schemes.* BIT **19** (1979), 204 – 213

P. HENRICI.

[1] *Discrete variable methods in ordinary differential equations.* John Wiley, New York 1962

K. HEUN.

[1] *Neue Methode zur approximativen Integration der Differentialgleichungen einer unabhängigen Veränderlichen.* Z. Math. Phys. **45** (1900), 23 – 38

P. HOFMANN.

[1] *Asymptotic expansions for the discretization error of boundary value problems for the Laplace equation in rectangular domains.* Numer. Math. **9** (1967), 302 – 322

J. HORN.

[1] *Über eine nichtlineare Differenzengleichung.* Jahresber. Dtsch. Math.-Ver. **26** (1918), 230 – 251

D. B. HUNTER.

[1] *Romberg's method for certain integrals involving a singularity.* BIT **7** (1967), 200 – 205

[2] *The numerical evaluation of Cauchy principal values of integrals by Romberg integration.* Numer. Math. **21** (1973), 185 – 192

D. B. HUNTER & H. V. SMITH.

[1] *The evaluation of Cauchy principal value integrals involving unknown poles.* BIT **29** (1989), 512 – 517

A. HURWITZ.

[1] *Über die Einführung der elementaren transzendenten Funktionen in der algebraischen Analysis.* Math. Annalen **70** (1911), 33 – 47

CH. HUYGENS.

[1] *De circuli magnitudine inventa,* Leiden 1654. A translation (German)

can be found in: F. RUDIO, *Archimedes, Huygens, Lambert, Legendre – Vier Abhandlungen über Kreismessung.* Teubner, Leipzig 1892, 83 – 131

D. JAGERMAN.

[1] *Investigation of a modified midpoint quadrature formula.* Math. Comp. **20** (1966), 79 – 89

K. JETTER.

[1] *Eine kurze Anmerkung zur Romberg-Integration.* Numer. Math. **45** (1984), 275 – 281

D. C. JOYCE.

[1] *Survey of extrapolation processes in numerical analysis.* SIAM Rev. **13** (1971), 435 – 488

S. KARLIN.

[1] *Total positivity.* Stanford University Press, Stanford 1968

S. KARLIN & W. J. STUDDEN.

[1] *Tchebycheff systems: With applications in analysis and statistics.* Interscience, New York 1966

R. F. KING.

[1] *An extrapolation method of order four for linear sequences.* SIAM J. Numer. Anal. **16** (1979), 719 – 725

K. KNOPP.

[1] *Infinite sequences and series.* Dover Publications, New York 1956

K. KOMMERELL.

[1] *Das Grenzgebiet der elementaren und höheren Mathematik.* Köhler, Leipzig 1936

W. KUTTA.

[1] *Beitrag zur näherungsweisen Integration totaler Differentialgleichungen.* Z. Math. Phys. **46** (1901), 435 – 453

J. D. LAMBERT.

[1] *On the local error and the local truncation error of linear multistep methods.* BIT **30** (1990), 673 – 681

[2] *Numerical methods for ordinary differential systems.* John Wiley and Sons, Chichester 1991

E. LANDAU.

[1] *Vorlesungen über Zahlentheorie.* Hirzel, Stuttgart 1927

P.-J. LAURENT.

[1] *Un théoreme de convergence pour le procédé d'extrapolation de Richardson.* C. R. Acad. Sci. Paris **256** (1963), 1435 – 1437

[2] *Etude de procédés d'extrapolation en analyse numérique.* Thesis, University of Grenoble 1964

D. P. LAURIE.

[1] *Propagation of initial rounding error in Romberg-like quadrature.* BIT **15** (1975), 277 - 282

D. LEVIN & A. SIDI.

[1] *Two new classes of non-linear transformations for accelerating the convergence of infinite integrals and series.* Appl. Math. Comp. **9** (1981), 175 – 215

M. J. LIGHTHILL.

[1] *Introduction to Fourier analysis and generalised functions.* Cambridge Univ. Press, Cambridge 1959

CH. LUBICH.

[1] h^2 *-Extrapolation methods for differential-algebraic systems of index 2.* IMPACT Comp. Sc. Eng. **1** (1989), 260 – 268

[2] *Linearly implicit extrapolation methods for differential-algebraic systems.* Numer. Math. **55** (1989), 197 – 211

S. LUBKIN.

[1] *A method for summing infinite series.* J. Res. Nat. Bur. Standards **48** (1952), 228 – 254

T. LUE.

[1] *Asymptotic expansion and extrapolation for the finite element approximation of nonlinear elliptic partial differential equations* (Chinese). Math. Numer. Sin. **9** (1987), 194 – 199

J. N. LYNESS.

[1] *Symmetric rules for hypercubes, I, II, III.* Math. Comp. **19** (1965), 260 – 276, 394 – 407, 625 – 637

[2] *An error functional expansion for* N *-dimensional quadrature with an integrand function singular at a point.* Math. Comp. **30** (1976), 1 – 23

[3] *Quadrature over a simplex: Part 1. A representation for the integrand function.* SIAM J. Numer. Anal. **15** (1978), 122 – 133

[4] *Quadrature over a simplex: Part 2. A representation for the error functional.* SIAM J. Numer. Anal. **15** (1978), 870 – 887

[5] *The Euler-Maclaurin expansion for the Cauchy principal value integral.* Numer. Math. **46** (1985), 611 – 622

[6] *Quadrature over curved surfaces by extrapolation.* Math. Comp. **63** (1994), 727 – 740

J. N. LYNESS & L. M. DELVES.

[1] *On numerical contour integration round a closed contour.* Math. Comp. **21** (1967), 561 – 577

J. N. LYNESS & E. DE DONCKER.

[1] *On quadrature error expansions. Part I.* J. Comp. Appl. Math. **17** (1987), 131 – 149

[2] *Quadrature error expansions - Part II. The full corner singularity.* Numer. Math. **64** (1993), 355 – 370

J. N. LYNESS & B. J. MCHUGH.

[1] *Integration over multidimensional hypercubes. A progressive procedure.* Computer J. **6** (1963), 264 – 270

[2] *On the remainder term in the n -dimensional Euler-Maclaurin expansion.* Numer. Math. **15** (1970), 333 – 344

J. N. LYNESS & G. MONEGATO.

[1] *Quadrature error expansions for the simplex when the integrand function has singularities at vertices.* Math. Comp. **34** (1980), 213 – 225

J. N. LYNESS & B. W. NINHAM.

[1] *Numerical quadrature and asymptotic expansions.* Math. Comp. **21** (1967), 162 – 178

J. N. LYNESS & K. K. PURI.

[1] *The Euler-Maclaurin expansion for the simplex.* Math. Comp. **27** (1973), 273 – 293

G. I. MARCHUK & V. V. SHAIDUROV.

[1] *Difference methods and their extrapolations.* Springer, New York 1983

A. C. MATOS.

[1] *Extrapolation algorithms based on the asymptotic expansion of the inverse of the error; application to continued fractions.* J. Comp. Appl. Math. **32** (1990), 179 – 190

G. MEINARDUS.

[1] *Approximation of functions: Theory and numerical methods.* Springer, Berlin/Heidelberg/New York 1967

[2] *Schnelle Fourier-Transformation.* In: *Numerische Methoden der Approximationstheorie* **4** (ISNM 42), L. COLLATZ, H. WERNER, G. MEINARDUS (eds.), Birkhäuser, Basel 1978, 192 – 203

[3] *On the asymptotic behavior of iteration sequences.* Proc. Fifth South African Symp. Numer. Anal., Durban 1979

[4] *Über das asymptotische Verhalten von Iterationsfolgen.* Z. Ang. Math. Mech. **63** (1983), 70 – 72

[5] *Private communication,* 1988

G. MEINARDUS & G. MERZ.

[1] *Praktische Mathematik I.* Bibl. Institut, Mannheim/Wien/Zürich 1979

[2] *Praktische Mathematik II.* Bibl. Institut, Mannheim/Wien/Zürich 1982

G. MEINARDUS & G. WALZ.

[1] *On the zeros of Cauchy polynomials with application to optimal depreciation.* J. Math. Econ. **20** (1991), 411 – 427

[2] *Approximation theory.* Springer, Berlin/Heidelberg/New York, in preparation

A. MEIR & A. SHARMA.

[1] *On the method of Romberg quadrature.* SIAM J. Numer. Anal. **2** (1965), 250 – 258

M. MEISTER.

[1] *Eine asymptotische Entwicklung des globalen Diskretisierungsfehlers von Differenzenverfahren zur numerischen Lösung von quasilinearen parabolischen Systemen von Differentialgleichungen.* Dissertation, TH Darmstadt 1988

[2] *An asymptotic expansion of the global discretization error of difference schemes for numerically solving a quasilinear parabolic system of differential equations.* Computing **47** (1992), 295 – 308

A. MEYERS.

[1] *Konvergenzverbesserung bei skalaren Iterationen.* Z. Ang. Math. Mech. **66** (1986), T419 – T421

C. A. MICCHELLI.

[1] *On a numerically efficient method for computing multivariate B-splines.* In: *Multivariate Approximation Theory,* W. SCHEMPP & K. ZELLER (eds.), Birkhäuser, Basel 1979, 211 – 248

[2] *A constructive approach to Kergin interpolation in $\mathbb{R}^k$: Multivariate B-splines and Lagrange interpolation.* Rocky Mountain J. Math. **10** (1980), 485 – 497

J. C. MILLER.

[1] *Neville's and Romberg's processes: A fresh appraisal with extensions.* Phil. Trans. Royal Soc. London, Ser. A **263** (1968), 525 – 562

C. MOLER & C. VANLOAN.

[1] *Nineteen dubious ways to compute the exponential of a matrix.* SIAM Rev. **20** (1978), 801 – 836

G. MONEGATO & J. N. LYNESS.

[1] *On the numerical evaluation of a particular singular two-dimensional integral.* Math. Comp. **33** (1979), 993 – 1002

G. MÜHLBACH.

[1] *A recurrence formula for generalized divided differences and some applications.* J. Approx. Theory **9** (1973), 165 – 172

[2] *Newton- und Hermite-Interpolation mit Chebysev-Systemen.* Z. Ang. Math. Mech. **54** (1974), 541 – 550

[3] *The general Neville-Aitken-algorithm and some applications.* Numer. Math. **31** (1978), 97 – 110

[4] *On two general algorithms for extrapolation with applications to numerical differentiation and integration.* In: *Padé approximation and its applications* (Lecture Notes Math. 888), H. BRASS & G. HÄMMERLIN (eds.), Springer, Berlin/New York 1981, 326 – 340

I. NAVOT.

[1] *An extension of the Euler-Maclaurin summation formula to functions with a branch singularity.* J. Math. Phys. **40** (1961), 271 –276

[2] *A further extension of the Euler-Maclaurin summation formula.* J. Math. Phys. **41** (1962), 155 –163

[3] *The Euler-Maclaurin functional for functions with a quasi-step discontinuity.* Math. Comp. **17** (1963), 337 – 345

[4] *The Euler-Maclaurin functional for functions with a complex singularity near the range of integration.* SIAM J. Numer. Anal. **2** (1965), 259 – 264

M. NEAMTU.

[1] *Multivariate Splines.* Doctoral Thesis, Enschede 1991

[2] *Multivariate divided differences I: Basic properties.* SIAM J. Numer. Anal. **29** (1992), 1435 – 1445

H. NEUMANN.

[1] *Über Fehlerabschätzungen zum Romberg-Verfahren.* Z. Ang. Math. Mech. **46** (1966), 152 – 153

B. W. NINHAM.

[1] *Generalized functions and divergent integrals.* Numer. Math. **8** (1966), 444 – 457

B. W. NINHAM & J. N. LYNESS.

[1] *Further asymptotic expansions for the error functional.* Math. Comp. **23** (1969), 71 – 83

N. E. NÖRLUND.

[1] *Vorlesungen über Differenzenrechnung.* Chelsea Publ., New York 1954

F. W. Olver.
[1] *Asymptotics and special functions.* Academic Press, New York 1974

G. Opfer.
[1] *On monotonicity preserving linear extrapolation sequences.* Computing **5** (1970), 259 – 266

[2] *On the monotonicity preservation of polynomial extrapolation sequences.* Appl. Anal. **1** (1971), 283 – 295

[3] *Monotonieerhaltung bei Polynomextrapolation.* Z. Ang. Math. Mech. **52** (1972), T214 – T216

H. Ortloff.
[1] *Asymptotisches Verhalten und Konvergenzbeschleunigung von Iterationsfolgen.* Numer. Math. **49** (1986), 545 – 559

[2] *Konvergenzbeschleunigung iterativer Algorithmen durch Extrapolationsmethoden.* Dissertation A, TH Magdeburg 1983

N. Osada.
[1] *A convergence acceleration method for some logarithmically convergent sequences.* SIAM J. Numer. Anal. **27** (1990), 178 – 189

A. Ostermann.
[1] *A half-explict extrapolation method for differential-algebraic systems of index 3.* IMA J. Numer. Anal. **10** (1990), 171 – 180

S. Paszkowski.
[1] *Convergence acceleration of continued fractions of Poincare's type 1.* Numer. Algorithms **2** (1992), 155 – 170

V. Pereyra.
[1] *Accelerating the convergence of discretization algorithms.* SIAM J. Numer. Anal. **4** (1967), 508 – 533

T. Popoviciu.
[1] *Sur le reste dans certaines formules linéaires d'approximation de l'analyse.* Mathematica (Cluj) **1** (1959), 95 – 142

Ph. Rabinowitz.
[1] *Extrapolation methods in numerical integration.* Numer. Algorithms **3** (1992), 17 – 28

M. Reimer.
[1] *Finite difference forms containing derivatives of higher order.* SIAM J. Numer. Anal. **5** (1968), 725 – 738

R. Remmert.
[1] *Funktionentheorie II.* Springer, Berlin/Heidelberg/New York 1991

L. F. Richardson.

[1] *The approximate arithmetical solution by finite differences of physical problems including differential equations, with an application to the stresses in a masonry dam.* Phil. Trans. Royal Soc. London, Ser. A **210** (1910), 307 – 357

[2] *How to solve differential equations approximately by arithmetic.* Math. Gazette **12** (1925), 415 – 421

[3] *The deferred approach to the limit. Part I: Single lattice.* Phil. Trans. Royal Soc. London, Ser. A **226** (1927), 299 – 349

B. Riemann.

[1] *Über die Anzahl der Primzahlen unter einer gegebenen Größe.* Monatsberichte der Berliner Akademie (1859), 671 – 680

W. Romberg.

[1] *Vereinfachte Numerische Integration.* Det Kong. Norske Vid. Selskab Forhdl. **28** (1955), 30 – 36

C. S. Rudisill & Y. Y. Chu.

[1] *Numerical methods for evaluating the derivatives of eigenvalues and eigenvectors.* AIAA J. **13** (1975), 834 – 837

C. Runge.

[1] *Über die numerische Auflösung von Differentialgleichungen.* Math. Ann. **46** (1895), 167 – 178

[2] *Über die numerische Auflösung totaler Differentialgleichungen.* Nachr. Ges. Wiss. Göttingen, Math.-Phys. Kl. (1905), 252 – 257

H. Rutishauser.

[1] *Ausdehnung des Rombergschen Prinzips.* Numer. Math. **5** (1963), 48 – 54

H. Rutishauser & E. Stiefel.

[1] *Remarques concernant l'intégration numérique.* Compt. Rend. Acad. Sci. Paris **252** (1961), 1899 – 1900

T. L. Saaty.

[1] *Modern Nonlinear Equations.* McGraw Hill, New York 1967

P. Sablonierre.

[1] *Asymptotic behaviour of iterated modified* Δ^2 *and* θ_2 *transforms on some slowly convergent sequences.* Numer. Algorithms **3** (1992), 401 – 409

J. F. Saigey.

[1] *Problèmes d'arithmétique et exercises de cacul du second degré avec les solutions raisonnées.* L. Hachette, Paris 1859

H. E. Salzer.

[1] *A simple method for summing certain slowly convergent series.* J. Math. Phys. **33** (1954), 356 – 359

[2] *Divided differences for functions of two variables for irregularly spaced points.* Numer. Math. **6** (1964), 68 – 77

G. Sansone & J. Geretsen.

[1] *Lectures on the theory of functions of a complex variable.* Noordhoff, Groningen 1960

J. W. Schmidt.

[1] *Asymptotische Einschließung bei konvergenzbeschleunigenden Verfahren.* Numer. Math. **8** (1966), 105 – 113

[2] *Asymptotische Einschließung bei konvergenzbeschleunigenden Verfahren II.* Numer. Math. **11** (1968), 53 – 56

C. Schneider.

[1] *Vereinfachte Rekursionen zur Richardson-Extrapolation in Spezialfällen.* Numer. Math. **24** (1975), 177 – 184

[2] *Error analysis for numerical integration - an algorithmic approach.* Habilitationsschrift, Mainz 1983.

A. Schönhage.

[1] *Mehrdimensionale Romberg-Integration.* Numer. Math. **14** (1970), 299 – 304

L. F. Shampine.

[1] *Efficient extrapolation methods for ODEs.* IMA J. Numer. Anal. **3** (1983), 383 – 395

[2] *Efficient extrapolation methods for ODEs II.* IMA J. Numer. Anal. **5** (1985), 23 – 28

[3] *Control of step size and order in extrapolation codes.* J. Comp. Appl. Maths. **18** (1987), 3 – 16

L. F. Shampine & L. S. Baca.

[1] *Smoothing the extrapolated midpoint rule.* Numer. Math. **41** (1983), 165 – 175

[2] *Global error estimates for odes based on extrapolation methods.* SIAM J. Sci. Stat. Comp. **6** (1985), 1 – 14

D. Shanks.

[1] *Nonlinear transformations of divergent and slowly convergent series.* J. Math. Phys. **34** (1955), 1 – 42

J. A. SHANKS.

[1] *On forming the Romberg table.* J. Comp. Appl. Math. **11** (1984), 343 - 351

A. SIDI.

[1] *Euler-Maclaurin expansions for integrals over triangles and squares of functions having algebraic/logarithmic singularities along an edge.* J. Approx. Theory **39** (1983), 39 - 53

[2] *Convergence and stability properties of minimal polynomial and reduced rank extrapolation algorithms.* SIAM J. Num. Anal. **23** (1986), 197 - 209

[3] *Generalizations of Richardson extrapolation with applications to numerical integration.* In: *Numerical integration III* (ISNM 85), H. BRASS & G. HÄMMERLIN (eds.), Birkhäuser, Basel 1988, 237 - 250

[4] *Computation of oscillary infinite integrals by extrapolation methods.* In: *Numerical integration,* T. O. ESPELID & A. GENZ (eds.), Kluwer Academic, Dordrecht 1992, 349 - 351

A. SIDI & J. BRIDGER.

[1] *Convergence and stability analyses for some vector extrapolation methods in the presence of defective iteration matrices.* J. Comp. Appl. Math. **22** (1988), 35 - 61

R. SKEEL.

[1] *Analysis of fixed-stepsize methods.* SIAM J. Numer. Anal. **13** (1976), 664 - 685

G. STARIUS.

[1] *Asymptotic expansions for a class of elliptic difference schemes.* Math. Comp. **37** (1981), 321 - 326

F. STETTER.

[1] *On a generalization of the midpoint rule.* Math. Comp. **22** (1968), 661 - 663

H. J. STETTER.

[1] *Asymptotic expansions for the error of discretization algorithms for non-linear functional equations.* Numer. Math. **7** (1965), 18 - 31

[2] *Symmetric two-step algorithms for ordinary differential equations.* Computing **5** (1970), 267 - 280

[3] *Analysis of discretization methods for ordinary differential equations.* Springer, Berlin 1973

E. STICKEL.

[1] *Über die Berechnung elementarer Matrixfunktionen.* Dissertation, University of Ulm 1985

E. STIEFEL.

[1] *Altes und Neues über numerische Quadratur.* Z. Ang. Math. Mech. **41** (1961), 408 – 413

J. STOER.

[1] *Über zwei Algorithmen zur Interpolation mit rationalen Funktionen.* Numer. Math. **3** (1961), 285 – 304

[2] *Extrapolation methods for the solution of initial value problems and their practical realization.* In: *Proc. Conf. Numer. Solution of ODEs* (Lecture Notes Math. 362), D. G. BETTIS (ed.), Springer, Berlin/Heidelberg/ New York 1974, 1 – 21

J. STOER & R. BULIRSCH.

[1] *Introduction to numerical analysis* (2nd ed.). Springer, Berlin/ Heidelberg/New York 1993

T. STRÖM.

[1] *Strict error bounds in Romberg quadrature.* BIT **7** (1967), 314 – 321

[2] *Monotonicity in Romberg quadrature.* Math. Comp. **26** (1972), 461 – 465

[3] *Monotonicity and error bounds in schemes of Romberg type for the computation of $f^{(n)}(0)$ by central differences and extrapolation.* Numer. Math. **21** (1973), 14 – 21

T. STRÖM & J. N. LYNESS.

[1] *On numerical differentiation.* BIT **15** (1975), 314 – 322

W. SWELDENS & R. PIESSENS.

[1] *Quadrature formulae and asymptotic error expansions for wavelet approximations of smooth functions.* SIAM J. Numer. Anal. **31** (1994), 1240 – 1264

[2] *Asymptotic error expansion of wavelet approximations of smooth functions II.* Numer. Math. **68** (1994), 377 – 401

R. C. TAN.

[1] *Some acceleration methods for iterative computation of derivatives of eigenvalues and eigenvectors.* Int. J. Numer. Meth. Eng. **28** (1989), 1505 – 1519

G. F. THOMAS.

[1] *On the implementation and performance of Walz's algorithm for the matrix exponential.* Research Report No. 90-205-K, Ontario Hydro 1990

R. S. VARGA & A. J. CARPENTER.

[1] *On the Bernstein Conjecture in approximation theory.* Constr. Approx. **1** (1985), 333 – 348

M. VAN VELDHUIZEN.

[1] *Asymptotic expansions of the global error for the implicit midpoint rule (stiff case).* Computing **33** (1984), 185 – 192

P. VERLINDEN & R. COOLS.

[1] *Proof of a conjectured asymptotic expansion for the approximation of surface integrals.* Math. Comp. **63** (1994), 717 – 725

P. VERLINDEN & A. HAEGEMANS.

[1] *An asymptotic expansion in wavelet analysis and its application to accurate numerical wavelet decomposition.* Numer. Algorithms **2** (1992), 287 – 298

R. DE VOGELAERE.

[1] *On a paper of Gaunt concerned with the start of numerical solutions of differential equations.* Z. Ang. Math. Phys. **8** (1957), 151 – 156

G. WALZ.

[1] *Approximation von Funktionen durch asymptotische Entwicklungen und Eliminationsprozeduren.* Dissertation, University of Mannheim 1987

[2] *Computing the matrix exponential and other matrix functions.* J. Comp. Appl. Maths. **21** (1988), 119 – 123

[3] *Effiziente Berechnung spezieller Funktionen mittels asymptotischer Entwicklungen und Eliminationsprozeduren.* J. Approx. Theory **53** (1988), 198 – 220

[4] *Error bounds and stopping rules for extrapolation methods.* IMA J. Numer. Anal. **9** (1989), 185 – 198

[5] *A contour integral representation of linear extrapolation methods.* Numer. Math. **55** (1989), 477 – 480

[6] *Increasing the convergence modulus of an asymptotic expansion.* Appl. Num. Maths. **6** (1989/90), 415 – 423

[7] *Increasing the convergence modulus of an asymptotic expansion: An algorithm for numerical differentiation.* In: *Algorithms for approximation II,* J. C. MASON & M. G. COX (eds.), Chapman and Hall, London 1990, 387-394

[8] *On the solution of Beltrami's equation by complex planar splines.* Complex Variables **15** (1990), 65 – 74

[9] *Spline-Funktionen im Komplexen.* B. I.-Wissenschaftsverlag, Mannheim/Wien/Zürich 1991

[10] *Asymptotic expansions and acceleration of convergence for higher order iteration processes.* Numer. Math. **59** (1991), 529 – 540

[11] *The history of extrapolation methods in numerical analysis.* Mannheimer Math. Manuskripte **130**, 1991

[12] *Generalized divided differences, with applications to generalized B-splines.* Calcolo **29** (1992), 111 – 123.

[13] *Asymptotic expansions for classical and generalized divided differences including applications.* Numer. Algorithms **7** (1994), 161 – 171

J. Wang.

[1] *Asymptotic expansions and L^∞ -error estimates for mixed finite element methods for second order elliptic problems.* Numer. Math. **55** (1989), 401 – 430

[2] *Superconvergence and extrapolation for mixed finite element methods on rectangular domains.* Math. Comp. **56** (1991), 477 – 503

G. Wanner.

[1] *Runge-Kutta methods with expansions in even powers of h.* Computing **11** (1973), 81 – 85

P. C. Waterman, J. M. Yos & R. J. Abodeely.

[1] *Numerical integration of non-analytic functions.* J. Math. Phys. **43** (1964), 45 – 50

K. Weierstrass.

[1] *Zur Theorie der eindeutigen analytischen Functionen.* Math. Abhandl. Kgl. Akad. Wiss. Berlin (1876), 11 – 60

[2] *Formeln und Lehrsätze zum Gebrauch der elliptischen Funktionen.* Physica-Verlag, Würzburg 1962 (Reprint of the 2nd ed. 1893)

H. Werner.

[1] *Praktische Mathematik I* (3rd ed.). Springer, Berlin/Heidelberg/New York 1982

E. T. Whittaker & G. N. Watson.

[1] *A course in modern analysis* (4rd ed.). Cambridge Univ. Press, Cambridge 1965

J. Wimp.

[1] *Sequence transformations and their applications.* Academic Press, New York 1981

L. Wuytack.

[1] *A new technique for rational extrapolation to the limit.* Numer. Math. **17** (1971), 215 – 221

P. Wynn.

[1] *A note on Salzer's method for summing certain slowly convergent series.* J. Math. Phys. **35** (1956), 318 – 320

[2] *On a device for computing the $e_m(S_n)$ transformation.* Math. Tabl. Aids Comp. **10** (1956), 91 – 96

Index

74 Inverse Problems: Principles and Applications in Geophysics, Technology, and Medicine
Proceedings of the International Conference held in Potsdam, August 30–September 3, 1993
Edited by Gottfried Anger, Rudolf Gorenflo, Horst Jochmann, Helmut Moritz, and Wigor Webers
328 pages, 1993
ISBN 3-05-501579-7

75 Generalized Diffusion Operators
By Jörg-Uwe Löbus
115 pages, 1993
ISBN 3-05-501617-3

76 The Stokes Equations
By Werner Varnhorn
154 pages, 1994
ISBN 3-05-501634-3

77/79 Systems and Networks: Mathematical Theory and Applications
Proceedings of the International Symposium MTNS '93 held in Regensburg, Germany, August 2–6, 1993
Edited by Uwe Helmke, Reinhard Mennicken, and Josef Saurer
Volume I: Key Invited Lectures
410 pages, 1994
ISBN 3-05-501573-8
Volume II: Invited and Contributed Papers
930 pages, 1994
ISBN 3-05-501661-0
Set-ISBN 3-05-501676-9

78 The Stokes System in an Infinite Cone
By Paul Deuring
267 pages, 1994
ISBN 3-05-501639-4

80 Nonlinear Waves in Networks
By Felix Ali Mehmeti
171 pages, 1994
ISBN 3-05-501640-8

81 Parcella '94
Proceedings of the VI. International Workshop on Parallel Processing by Cellular Automata and Arrays held in Potsdam, September, 21–23, 1994
Edited by Chris Jesshope, Vesselin Jossifov, and Wolfgang Wilhelmi
347 pages, 1994
ISBN 3-05-501602-5

82 Partial Differential Equations
Models in Physics and Biology
Edited by Günter Lumer, Serge Nicaise, and Bert-Wolfgang Schulze
421 pages, 1994
ISBN 3-05-501657-2

83 Selberg Zeta and Theta Functions
A Differential Operator Approach
By Ulrich Bunke and Martin Olbrich
168 pages, 1995
ISBN 3-05-501690-4

84 Non-Linear Hyperbolic Equations in Domains with Conical Points
Existence and Regularity of Solutions
By Ingo Witt
230 pages, 1995
ISBN 3-05-501691-2

85 Stochastic Evolution Equations
By Wilfried Grecksch and Constantin Tudor
178 pages, 1995
ISBN 3-05-501697-1

86 Approximation Theory
Proceedings of the International Dortmund Meeting IDoMAT 95 held in Witten, Germany, March 13–17, 1995
Edited by Manfred W. Müller, Michael Felten, and Detlef H. Mache
332 pages, 1995
ISBN 3-05-501673-4